☑ Y0-ACH-667

This report contains the collective views of an
international group of experts and does not
necessarily represent the decisions or the stated
policy of either the World Health Organization,
the United Nations Environment Programme,
or the International Radiation Protection Association.

Environmental Health Criteria 22

ULTRASOUND

Published under the joint sponsorship of
the United Nations Environment Programme,
the World Health Organization, and the
International Radiation Protection
Association

World Health Organization
Geneva, 1982

PRINTED IN FINLAND

83/5631 – VAMMALA – 7000

CONTENTS

NOTE TO READERS OF THE CRITERIA DOCUMENTS

While every effort has been made to present information in the criteria documents as accurately as possible without unduly delaying their publication, mistakes might have occurred and are likely to occur in the future. In the interest of all users of the environmental health criteria documents, readers are kindly requested to communicate any errors found to the Division of Environmental Health, World Health Organization, Geneva, Switzerland, in order that they may be included in corrigenda which will appear in subsequent volumes.

In addition, experts in any particular field dealt with in the criteria documents are kindly requested to make available to the WHO Secretariat any important published information that may have inadvertently been omitted and which may change the evaluation of health risks from exposure to the environmental agent under examination, so that the information may be considered in the event of updating and re-evaluation of the conclusions in the criteria documents.

WHO/IRPA TASK GROUP ON ENVIRONMENTAL HEALTH CRITERIA FOR ULTRASOUND

Members

Dr V. B. Bindal, National Physical Laboratory, New Delhi, India

Dr P. D. Edmonds, Ultrasonic Program, Stanford Research Institute, Menlo Park, California, USA

Dr D. Harder, Institute for Medical Physics and Biophysics, University of Gottingen, Federal Republic of Germany [a]

Dr K. Lindström, Department of Biomedical Engineering, University Hospital, Malmö, Sweden

Dr K. Maeda, Department of Obstetrics and Gynaecology, Tottori University School of Medicine, Yonago, Japan

Dr V. Mazzeo, Department of Ophthalmology, University of Ferrara, Ferrara, Italy (Vice-Chairman)

Dr W. Nyborg, Department of Physics, University of Vermont, Burlington, Vermont, USA

Dr M. H. Repacholi, Radiation Protection Bureau, Department of National Health and Welfare, Ottawa, Canada (Chairman) [a]

Dr H. F. Stewart, Bureau of Radiological Health, Department of Health and Human Services, Food and Drug Administration, Rockville, Maryland, USA (Rapporteur)

Dr M. Stratmeyer, Bureau of Radiological Health, Department of Health and Human Services, Food and Drug Administration, Rockville, Maryland, USA (Rapporteur)

Dr A. R. Williams, Department of Medical Biophysics, University of Manchester, Manchester, United Kingdom, (Rapporteur)

Representatives of other Organizations

Dr W. D. O'Brien, American Institute of Ultrasound in Medicine, Department of Electrical Engineering, University of Illinois Urbana, Champaign, Illinois, USA

Dr C. Pinnagoda, International Labour Office, Geneva, Switzerland

Secretariat

Mrs A. Duchêne, Commissariat à l'Energie Atomique, Departement de Protection, Fontenay-aux-Roses, France [a]

Dr E. Komarov, Scientist, Environmental Hazards and Food Protection, Division of Environmental Health, World Health Organization, Geneva, Switzerland (Secretary)

----

[a] Members of the International Non-Ionizing Radiation Committee of IRPA

ENVIRONMENTAL HEALTH CRITERIA FOR ULTRASOUND

Further to the recommendations of the Stockholm United Nations Conference on the Human Environment in 1972, and in response to a number of World Health Assembly resolutions (WHA23.60, WHA24.47, WHA25.58, WHA26.68) and the recommendation of the Governing Council of the United Nations Environment Programme, (UNEP/GC/10, 3 July 1973), a programme on the integrated assessment of the health effects of environmental pollution was initiated in 1973. The programme, known as the WHO Environmental Health Criteria Programme, has been implemented with the support of the Environment Fund of the United Nations Environment Programme. In 1980, the Environmental Health Criteria Programme was incorporated into the International Programme on Chemical Safety (IPCS). The result of the Environmental Health Criteria Programme is a series of criteria documents.

A joint WHO/IRPA Task Group on Environmental Health Criteria for Ultrasound met in Geneva from 7 to 11 June 1982. Mr G. Ozolins, Manager, Environmental Hazards and Food Protection, opened the meeting on behalf of the Director-General. The Task Group reviewed and revised the draft criteria document, made an evaluation of the health risks of exposure to ultrasound, and considered rationales for the development of equipment performance standards and human exposure limits.

The International Radiation Protection Association (IRPA) became responsible for activities concerned with non-ionizing radiation by forming a Working Group on Non-Ionizing Radiation in 1974. This Working Group later became the International Non-Ionizing Radiation Committee (IRPA/INIRC) at the IRPA meeting in Paris in 1977. The IRPA/INIRC reviews the scientific literature on non-ionizing radiation and makes assessments of the health risks of human exposure to such radiation. Based on the Health Criteria Documents developed in conjunction with WHO, the IRPA/INIRC recommends guidelines on exposure limits, drafts codes of safe practice, and works in conjunction with other international organizations to promote safety and standardization in the non-ionizing radiation field.

Two WHO Collaborating Centres, the Radiation Protection Bureau, Health and Welfare Canada, and the Bureau of Radiological Health, Rockville, USA, cooperated with the IRPA/INIRC in initiating the preparation of the criteria document. The final draft was prepared as a result of several working group meetings, taking into account comments received from independent experts and the national focal points for the WHO Environmental Health Criteria Programme in Australia, Canada, Finland, Federal Republic of Germany, Italy, Japan, New Zealand, Sweden, the United Kingdom, the USA, and the USSR

as well as from the United Nations Environment Programme, the
Food and Agriculture Organization of the United Nations, and
the International Labour Organisation. The collaboration of
these experts, national institutions, and international
organizations is gratefully acknowledged. Without their
assistance this document could not have been completed. In
particular, the Secretariat wishes to express its thanks to Dr
D. Harder, Dr C. R. Hill, Dr M. H. Repacholi, Dr C. Roussell,
Dr H. F. Stewart, Dr M. E. Stratmeyer, and Dr A. R. Williams
for their assistance in the preparation of the draft document
and to Dr Repacholi and Dr Williams for their help in the
final scientific editing of the text.

The document is based primarily on original publications
listed in the reference section. Additional information was
obtained from a number of general reviews including: Nyborg,
(1977); Repacholi, (1981); and Stewart & Stratmeyer (1982).

Modern advances in science and technology change man's
environment, introducing new factors which, besides their
intended beneficial uses, may also have untoward side-effects.
Both the general public and health authorities are aware of
the dangers of pollution by chemicals, ionizing radiation, and
noise, and of the need to take appropriate steps for effective
control. The more frequent use of ultrasound in industry,
commerce, the home, and particularly in medicine, has
magnified the possibiity of human exposure, increasing concern
about possible human health effects, especially in relation to
the human fetus.

This document comprises a review of data, which are
concerned with the effects of ultrasound exposure on
biological systems, and are pertinent to the evaluation of
health risks for man. The purpose of this criteria document is
to provide information for health authorities and regulatory
agencies on the possible effects of ultrasound exposure on
human health and to give guidance on the assessment of risks
from medical, occupational, and general population exposure to
ultrasound.

Subjects briefly reviewed include: the physical
characteristics of ultrasound fields; measurement techniques;
sources and applications of ultrasound; levels of exposure
from devices in common use; mechanisms of interaction;
biological effects; and guidance on the development of
protective measures such as regulations or safe-use guidelines.

In a few countries, concern about occupational and public
health aspects has led to the development of radiation
protection guidelines and the establishment of equipment
emission or performance standards, and limits for human
exposure (mainly to airborne ultrasound). Health agencies and
regulatory authorities are encouraged to set up and develop

programmes which ensure that the lowest exposure occurs with the maximum benefit. It is hoped that this criteria document may provide useful information for the development of national protection measures against non-ionizing acoustic radiation.

Details of the WHO Environmental Health Criteria Programme, including definitions of some of the terms used in the documents, may be found in the general introduction to the Environmental Health Criteria Programme, published together with the environmental health criteria document on mercury (Environmental Health Criteria 1 - Mercury, Geneva, World Health Organization, 1976), now available as a reprint.

1. SUMMARY AND RECOMMENDATIONS FOR FURTHER STUDIES

1.1 Summary

1.1.1 Scope and purpose

This document comprises a review of data which are concerned with the effects of ultrasound exposure on biological systems and are pertinent to the evaluation of health risks for man. The purpose of this evaluation is to provide information for health authorities and regulatory agencies on the possible effects of ultrasound exposure on human health and to give guidance on the assessment of risks from medical, occupational, and general population exposure to ultrasound.

Subjects briefly reviewed include: the physical characteristics of ultrasound fields; measurement techniques; sources and applications of ultrasound; levels of exposure in common use; mechanisms of interaction; and guidance on the development of protective measures such as regulations or safe-use guidelines.

1.1.2 Introduction

Ultrasound is sound (a mechanical vibration phenomenon) having a frequency above the range of human hearing (typically above 16 kHz) which, unlike electromagnetic radiation, requires a medium through which to propagate.

Exposure to ultrasound can be divided into two distinct categories: airborne and liquid-borne. Exposure to airborne ultrasound occurs in many industrial applications such as cleaning, emulsifying, welding, and flaw detection and through the use of consumer devices such as dog whistles, bird and rodent controllers, and camera rangefinders, and commercial devices such as intrusion alarms. Liquid-borne exposure occurs predominantly through medical exposure in diagnosis, therapy, and surgery.

As with any other physical agent, ultrasound has the potential to produce adverse effects at sufficiently high doses. In addition, biological effects of unknown significance have been reported under laboratory conditions at low exposure levels. However, the health risks that may be associated with

biological effects at the levels of ultrasound currently
encountered in occupational, environmental, or medical
exposure have not yet been defined.

Though, at present, there is no evidence of adverse health
effects in human beings exposed to diagnostic ultrasound, its
rapidly increasing use during pregnancy is still of special
concern in view of the known susceptibility of the fetus to
other physical and chemical agents.

1.1.3 Mechanisms of action

Acoustic energy may be transformed into several other
forms of energy, which may exist at the same time within any
given medium. The mechanisms of transformation into these
other forms of energy are conventionally subdivided into three
major categories comprising a thermal mechanism, a
cavitational mechanism, and other mechanisms including
streaming motions.

When ultrasound is absorbed by matter, it is converted
into heat producing a temperature rise in the exposed subject.
An ultrasound wave produces alternate areas of compression and
rarefaction in the medium and the pressure changes produced
can result in cavitation. This phenomenon occurs when
expansion and contraction of nuclei or gas bubbles (in liquids
and body tissues) cause either simple oscillations or
pulsations (stable cavitation), or violent events (transient
or collapse cavitation), where the collapse of the bubbles
produces very high instantaneous temperatures and pressures.
Theoretical analyses have predicted that a single cycle of
ultrasound, at a sufficient amplitude level, can produce a
transient cavitation event in an aqueous medium in which
appropriate nucleation sites are present. This prediction has
not yet been verified experimentally.

Streaming motions and shearing stresses can occur within
the exposed system through stable cavitation; twisting motions
(radiation torque) have also been observed in biological
systems exposed to ultrasound.

Unlike ionizing radiation, where the basic physical
mechanism of interaction stays the same with increasing
exposure rate, the dominant mechanism of ultrasound action on
biological systems can change as the acoustic intensity,
frequency, and exposure conditions change.

It is generally agreed that diagnostic devices emitting space- and time-averaged intensities of the order of a few milliwatts/cm² are unlikely to cause temperature elevations in human beings that would be regarded as potentially damaging. It is not known whether some form of cavitational activity could occur in vivo at these time-averaged intensities when pulse-echo devices are used. It has been suggested that the elevated temperatures associated with the use of higher spatial average temporal average (SATA) intensities (0.1-3 W/cm²) contribute to the beneficial therapeutic effects of ultrasound. In addition, gas bubbles have been detected in vivo following therapeutic exposures, indicating that a form of cavitational activity has occurred.

1.1.4 Biological effects

Very few systematic studies have been undertaken to determine threshold levels for observed effects of ultrasound. Nearly all of the reports in the literature have tended to be phenomenological in nature, without evidence from further investigations to determine the underlying mechanisms of action. Furthermore, most reports have not yet been confirmed by more than one laboratory. Some studies have been performed using exposure times longer than would normally be encountered in the clinical situation and this has made the evaluation of health risks from exposure to ultrasound extremely difficult. Thus, there is an urgent need for more carefully coordinated, systematic research in critical areas.

The health implications from a number of effects already reported indicate the need for a prudent approach to the ultrasound exposure of human subjects, even though the benefits of this imaging modality far outweigh any presumed risks.

1.1.4.1 Airborne ultrasound

Exposure of human beings to low frequency ultrasound (16 - 100 kHz) can be divided into two distinct categories; one is via direct contact with a vibrating solid or through a liquid coupling medium, and the other is through airborne conduction.

For airborne ultrasound exposure, at least one of the critical organs is the ear. Effects reported in human subjects exposed to airborne ultrasound include; temporary threshold shifts in sound perception, altered blood sugar levels, electrolyte imbalance, fatigue, headaches, nausea, tinnitus, and irritability. However, in many instances, it has been difficult to state that the observed effects were caused by airborne ultrasound because they were subjective and there was often simultaneous exposure to high levels of audible sound.

The use of experimental animals to study the effects of airborne ultrasound has serious drawbacks because they have a greater hearing acuity, wider audible frequency range, and a greater surface-area-to-mass than man and most have fur-covered bodies.

1.1.4.2 Biological Molecules

Studies of the exposure of biological molecules in solution to liquid-borne ultrasound have, in general, served to indicate the importance of cavitation as a mechanism of ultrasound action and to identify which biological molecules preferentially absorb the energy. It is not possible to extrapolate the results of such studies to the in vivo situation.

1.1.4.3 Cells in suspension

There is evidence that ultrasound can change the rate of macromolecular synthesis and cause ultrastructural changes within cells. Alterations in cell membrane structure and function have been reported from exposure to pulsed and continous wave (cw) ultrasound using commercial diagnostic devices.

Conflicting results have been reported on the effects of ultrasound on DNA. Unscheduled DNA synthesis (indicating possible damage and subsequent repair to the DNA) has been reported following exposure to pulsed diagnostic ultrasound and cw therapeutic ultrasound.

Some evidence has been produced that alterations in cell surface activity may persist for many generations.

1.1.4.4 Organs and tissues

Studies on skeletal tissue indicate that bone growth may be retarded following exposure to ultrasound at high therapeutic intensities, even if the transducer is kept in motion during treatment. If the transducer is held stationary, bone and other tissue damage occurs at lower intensities.

Both in vitro and in vivo exposures of muscle tissue have been reported to trigger contractions. Therapeutic intensities of ultrasound have also been reported to alter thyroid function in man.

1.1.4.5 Animal studies

Fetal weight reduction has been observed following exposure of rodent fetuses in utero. The lowest cw average intensity levels that have been observed to induce fetal weight reduction in mice are in the low therapeutic range. Some studies indicate that fetal abnormalities and maternal weight loss also occur.

1.1.4.6 Epidemiology and health risk evaluation

To date, adverse effects have not been detected from exposure to diagnostic ultrasound. However, it is of particular concern that adequate epidemiological studies have not yet been performed, and that soon most human fetuses in technologically developed countries could be subjected to at least one ultrasound examination. If such epidemiological studies are not carried out very soon, there will not be any "control" populations to compare with populations exposed to ultrasound.

Most of the human studies that have been performed have suffered from inadequate control matching, too few cases, or a variety of other problems and though, in general, adverse effects have not been reported, these studies are inconclusive and of very little value. The possibility of reduced weight resulting from in utero exposure, which was reported recently, still needs further investigation, especially in light of previous reports of reduced body weight in animal fetuses exposed in utero.

1.1.5 Exposure limits and emission standards

1.1.5.1 Occupational exposure to airborne ultrasound

Occupational exposure limits for airborne ultrasound have already been established or have been proposed in Canada, Japan, Sweden, the United Kingdom, the USA, and the USSR. All standards or proposed standards or regulations are similar, in that each has a "step" allowing exposure to sound pressure levels above 20 kHz.[a]

1.1.5.2 Therapeutic use

Regulations which incorporate maximum output levels for therapeutic ultrasound equipment exist in some countries (e.g., Canada) and have been proposed as a requirement by one technical sub-committe of the International Electrotechnical Commission. Other countries, such as the USA, have not incorporated a limit on output levels in their ultrasound therapy products standard.

1.1.5.3 Diagnostic use

Given the current biological and biophysical data base, there does not appear to be sufficient information to establish quantitative limits on output levels for diagnostic ultrasound equipment.

1.1.5.4 General population exposure

Ultrasound is used in many consumer products (e.g., camera range-finders and TV controls, burglar alarms etc.) but little is known about their potential health effects in the general population, although they are thought to be negligible.

1.2 Recommendations for Further Studies

1.2.1 Measurement of ultrasonic fields

One of the difficulties of establishing a comprehensive body of information with respect to the biological and health effects of ultrasound has been the lack of adequate instrumentation to measure the various exposure parameters. However, reliable methods for the measurement of ultrasound

---

[a] The International Radiation Protection Association is proposing guidelines on limits of exposure to airborne acoustic energy for both workers and the general population.

field parameters, such as total radiated power, and the various intensities in the ultrasound fields, are now available in a few national or research institutions.

Most devices used to measure ultrasound power and the various temporal and spatial intensity parameters for liquid-borne ultrasound are not suitable for routine surveys in the work place. There is an urgent need for the development of portable, rugged instrumentation that will measure accurately both total power and the relevant intensity parameters.

Furthermore, a substantial research effort is still needed to develop a system of dosimetric variables relevant to the production of and protection against adverse health effects of ultrasound in medical and industrial applications.

1.2.2 Exposure of patients to diagnostic ultrasound

Information concerning the ultrasound exposure of patients during diagnostic examinations has often not been available in the past. Manufacturers are now increasingly supplying diagnostic ultrasound equipment together with appropriate data to enable users to evaluate the level to which the patient is exposed, and to decide which devices would give the lowest exposure commensurate with good diagnostic quality. This trend is commendable and should be strongly encouraged.

Until the potential health effects of exposure to ultrasound have been properly evaluated, it is recommended that manufacturers should aim at keeping the output levels necessary for examinations as low as readily achievable. This priority should apply to all diagnostic techniques where the exposure time required to conduct the examination can be minimized.

It is strongly recommended that patients should only be exposed to ultrasound for valid clinical reasons.

1.2.3 Biological studies

Most bioeffect studies have been conducted on cell suspensions, plants, insects, and other animal systems. However, it should be noted that some of these biological systems accentuate certain mechanisms of interaction to the extent that effects are observed under exposure conditions that would not apply to, or would not induce effects in human beings. Controversy continues as to the exact mechanisms by which the effects of ultrasound are induced. It is often possible to distinguish between dominant thermal and non-thermal mechanisms, but the type of non-thermal effect remains open to discussion. Cavitation is a well established mechanism of action, but other non-thermal mechanisms may be involved in the production of some ultrasound effects. With more complete information on biological and physical mechanisms, studies can be undertaken to determine possible thresholds (if they exist) for bioeffects and the biophysical knowledge could be used to predict potential bioeffects.

(a) Molecules and cells
It is recommended that studies be conducted at both the molecular and cellular levels on interactions between ultrasound and biological systems. Such information is needed to evaluate the importance of the interaction mechanisms involved and to clarify areas and end-points that need further study at higher levels of biological organization.

(b) Immunological studies
Recent studies suggest that ultrasound may induce immunological responses in laboratory animals. Because of the fundamental importance of the immune system, any effects that might be induced by ultrasound should be systematically investigated.

(c) Haematological studies
Ultrasound at therapeutic intensities has been shown to cause platelet aggregation and other haematological alterations in vitro. Results of some studies suggest that similar effects may occur in vivo. This suggestion needs to be investigated further to assess possible adverse consequences in vivo.

(d) Effects on DNA
Recent studies reporting repair to DNA, observed as unscheduled DNA synthesis, need to be substantiated. Of particular importance is the investigation of damage to DNA from pulsed ultrasound with intensities in the diagnostic range.

(e) Genetic effects
Reports of sister chromatid exchanges, increased transformation frequency, and changes in the cell membrane and cell motility, seen many generations after a single exposure to ultrasound, suggest a "genetic" effect. Because these results have not been adequately confirmed, they cannot, at present, be extrapolated to the in vivo situation; and need further investigation.

(f) Fetal studies

A number of reports indicate that lower fetal weight and increased fetal abnormalities occur following exposure to ultrasound in the low therapeutic intensity range. Studies should be undertaken to establish exposure thresholds (if any) for effects on the fetus exposed on various days during gestation. The importance of the ratio of temporal average to temporal peak intensities in relation to the production of fetal effects also needs considerable investigation.

Since gross effects appear to occur only at high ultrasound intensities, research workers should concentrate their efforts on subtle effects, particularly in the fetus, which in many instances receives a whole-body exposure to ultrasound. Wherever possible, studies should be related to clinical situations.

Only one study on human beings suggests that lower birthweights may result from exposure to diagnostic ultrasound in utero.

As the practice of ultrasound diagnosis becomes more widespread, it will be difficult to find adequate control populations and opportunities for satisfactory epidemiological studies may become increasingly rare. It is strongly recommended that cost-effective, well-designed studies be conducted soon and coordinated at both the national and international levels.

Short-term studies where specific end-points, such as haematological effects, can be identified, also need to be conducted. Investigations should be made on patients undergoing ultrasound therapy, since the average intensities used are significantly higher than those used in diagnosis. To date, such studies do not seem to have been undertaken.

(g) Behavioural studies

Studies on rodents suggest that behavioural effects may be seen in newborn that have been exposed in utero. If these studies are confirmed, systematic studies on human newborn will be necessary, to determine whether such effects occur in man.

(h) Synergism

It is common for patients to undergo diagnostic examinations, on the same day, in both the ultrasound and X-ray departments of hospitals. Some evidence has been produced indicating that X-rays may enhance ultrasound effects. Increased chromosome aberration rates in somatic cells have been observed following combined exposure to ultrasound and X-rays. Preliminary reports also suggest that ultrasound may have a synergistic action with such agents as heat, viruses, and drugs. Such synergistic effects need to be investigated further.

(i) Airborne ultrasound

Few studies have been reported on the effects of airborne ultrasound on man. Earlier reports of headaches and nausea seem to have been largely attributed to subharmonics of the ultrasound beam in the audible range. However, there has been a number of reports of similar symptoms from people exposed to devices such as ultrasound intrusion alarms. This indicates that further investigation in this area is necessary.

1.2.4 Training and education

Since the ultrasound exposure levels currently employed in physiotherapy are well within the range in which adverse health effects have been confirmed, it is recommended that all operators of such equipment receive formal training (up to 1 year) before treating patients. These operators should also ensure that their equipment is properly maintained and calibrated to make sure that patients receive only the prescribed "dose".

Operators of diagnostic ultrasound equipment should also receive appropriate formal training on the use and safety of this clinical modality. They should be properly instructed on maintaining and calibrating the equipment to ensure that the ultrasound exposure of the patient is minimized while maximizing the quality of the image.

In commercial, industrial, and research establishments where devices emitting airborne and/or liquid-borne ultrasound operate, all potentially exposed employees should be properly instructed with regard to safety precautions appropriate for the equipment being used.

Consumers using devices that emit airborne ultrasound should familiarize themselves with the safety precautions provided by the manufacturer.

1.2.5 Regulations and safety guidelines for equipment

Protective measures include the use of either mandatory standards (regulations) or guidelines on equipment emission and performance.

Where appropriate, safety guidelines should be provided for operators of equipment that emits airborne ultrasound. In many cases, such guidelines should recommend the use of hearing-protectors and appropriate warning signs.

As surveys indicate, many ultrasound therapy devices do not give the output levels indicated on the control console, so mandatory standards or regulations are recommended for this type of equipment. Such standards should include accuracy specifications for the output power, output intensity, and timer setting.

The establishment of guidelines on the performance of diagnostic ultrasound equipment is recommended and these should include requirements concerning the image quality and stability, and quality assurance measures. At present, there does not appear to be a need to limit the output exposure levels of diagnostic ultrasound equipment, other than to recommend strongly that the lowest output levels be used commensurate with image quality, adequate to obtain the necessary diagnostic information.

2. PHYSICAL CHARACTERISTICS OF ULTRASOUND

Ultrasonic energy consists of mechanical vibrations occurring above the upper frequency limit of human audibility (generally accepted as about 16 kHz). Ultrasound consists of a propagating disturbance in a medium, which causes subunits (particles) of the medium to vibrate. The vibratory motion of the particles characterizes ultrasonic (acoustic) energy propagation. Unlike electromagnetic radiation, acoustic energy cannot be transmitted through a vacuum. The transmission through the medium depends to a great extent on the ultrasound frequency and the state of the medium, i.e., gas, liquid, or solid.

Ultrasound may propagate in different modes. In solids, two important modes include compressional (longitudinal) waves and shear (transverse) waves (Fig. 1). The propagation velocities of these two modes are generally different.

Ultrasound propagates in gaseous, liquid, or solid media, mainly in the form of longitudinal or compressional waves formed by alternate regions of compression and rarefaction of the particles of the medium, which vibrate in the direction of energy propagation. The distance between two consecutive points of maximum compression or rarefaction is called the wavelength.

Transverse (shear) waves mainly propagate in solids, and are characterized by particle displacement at $90°$ to the direction of propagation. At a bone/soft tissue interface, one type of wave can give rise to another (mode conversion). If a longitudinal wave propagating in soft tissue strikes bone at an angle, both longitudinal and transverse waves may be excited in the solid medium. This phenomenon can result in heating at the bone surface. Results of heating in bone have been reported by Lehmann & Guy (1972) and Chan et al. (1974).

The passage of a sound wave through a medium can be characterized by several variables, associated with the movement of particles in the medium. These include: acoustic pressure (p), particle displacement (ξ), particle velocity (v), and particle acceleration (a). Under idealized conditions each of these quantities varies sinusoidally with space and time (Appendix I).

The acoustic pressure (p) is the change in total pressure at a given point in the medium at a given time, resulting in compression where p is positive, and expansion where p is negative, as a result of the action of the ultrasound waves. The displacement (ξ) is the difference between the mean position of a particle in the medium and its position at any given instant in the time (t). The particle velocity (v) is

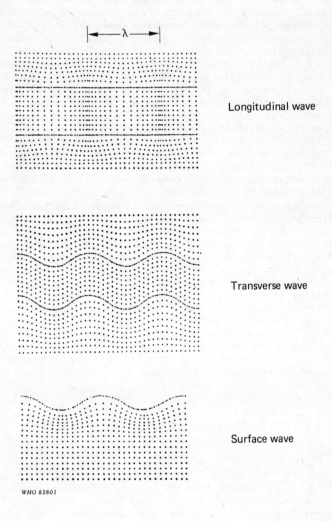

Longitudinal wave

Transverse wave

Surface wave

WHO 82801

Fig. 1. Propagation modes of ultrasound waves. (a) Compressional or longitudinal waves, (b) shear or transverse waves, illustrated by particle displacement from an undisturbed position. The direction of propagation is from left to right; conditions at which the boundaries of an ultrasound beam of limited width are indicated.

the instantaneous velocity of a vibrating particle at a given point in the medium. This should not be confused with the speed of sound ($\underline{c}$). The latter is the speed with which the wave propagates through the medium, even though the individual particles of the medium vibrate only about their mean positions with no bulk movement of matter. The speed of sound ($\underline{c}$) is a constant that depends on the physical properties of the medium; it is discussed in section 2.3. As a result of the sinusoidal variation in particle velocity ($\underline{v}$), each particle experiences an acceleration ($\underline{a}$) which also varies with time and position; it has positive values when v increases, and negative values when $\underline{v}$ decreases.

The relationship between the intensity and various particle parameters such as acoustic pressure, displacement, velocity, and acceleration (Appendix I, Table 1) may be of importance when analysing some biological effects reported in the literature.

For comparative purposes, it is worth noting an important difference between ionizing radiation and ultrasound. To increase the intensity of a beam of X-rays of a given spectral distribution, the photon flux is increased. The energy of each individual photon remains unchanged. Therefore, the inter-action mechanism for each photon remains the same, but the number of interactions per unit time increases because of the increased number of photons. To increase the intensity of a beam of ultrasound of fixed frequency, the amplitude of the particle parameters (pressure, displacement, velocity, accel-eration) is increased, to obtain a higher energy flux per unit area. Change in the magnitude of the particle parameters may affect the relative importance of different mechanisms of interaction with matter at different intensities.

2.1 Continuous, Gated, and Pulsed Waves

The differences between continuous wave, gated (amplitude-modulated), and acoustic-burst pulsed waves are shown in Fig. 2. A continuous wave at a single frequency is a simple sinusoidal wave having constant amplitude. Amplitude-modulated waveforms are used in some equipment, for example, pulsed therapy equipment. An acoustic burst is the type of pulse used in pulse echo diagnostic equipment. It can represent the variation of pressure as a function of distance at a fixed instant in time, or as a function of time at a fixed point in space. For the pulsed wave, the pressure amplitude is not constant and is zero for part of the time. No acoustic energy is being emitted between pulses and the ultrasound propagates

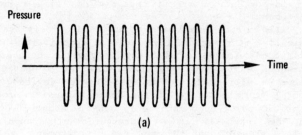

(a)

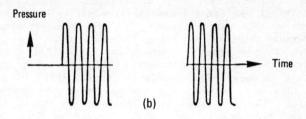

(b)

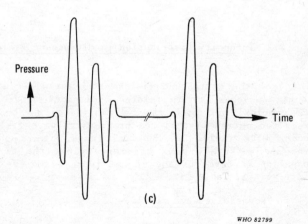

(c)

Fig. 2. Ultrasonic waves. (a) Continuous; (b) gated continuous; (c) pulsed.

through the medium as small packages of acoustic energy.
Pulsed waves can have any combination of on/off times. Thus,
it is important to specify exactly the time regimen of the
pulsed beam.

Pulsed ultrasound with short and widely-spaced pulses
(typically microsecond (μs) pulses spaced at intervals of
milliseconds (ms)) is used for diagnostic purposes, whereas
continuous waves (cw) are often used in therapeutic applications of ultrasound and in most Doppler devices. Though the
temporal (time) average of the sound intensity produced by a
diagnostic pulse echo machine is usually about 1000 times less
than the intensity in a therapeutic ultrasound beam, the
acoustic pressure and the particle displacement, velocity, and
acceleration during the pulse may reach peak values an order
of magnitude greater than those in cw therapeutic ultrasound.

A particularly complex time structure of the ultrasound
field may occur with real-time diagnostic devices that have an
array of transducers, where acoustic beams emitted by adjacent
elements of the array sequentially contribute to the acoustic
intensity at a point in space. The temporal characteristics of
ultrasound fields such as pulse duration, pulse repetition
frequency, and temporal peak intensity have been reported by
several investigators (Barnett, 1979; Child et al., 1980a;
Lewin & Chivers, 1980; Sarvazyan et al., 1980). A distinction
must be made between the spatial peak intensity and the
spatial average intensity (Appendix I, Table 1); great
differences between particle parameters can occur over space
as well as time. Considerable spatial variations in pressure
occur in a standing wave field (section 2.2.2).

2.2 Intensity Distribution in Ultrasound Fields

Many of the ultrasound fields encountered during exposure
of human subjects, or in related biological studies, may be
quite complex, but most can be considered to be somewhere
between two extreme types: the progressive wave field and the
standing wave field. In the first case, it is possible to
define and measure a flux of energy along the direction of
propagation in terms of any of the four parameters (p, ξ, v,
a) (Appendix I, Table 1).

2.2.1 Progressive wave fields

The ultrasonic field produced by a transducer obeys all
the physical laws of wave phenomena. It can be thought of as
being produced by many small point sources making up the

transducer face and thus producing a characteristic interference pattern at any point in the field. As ultrasound is propagated from the transducer, there is a zone where the overall beam size remains relatively constant (the near field), though there are many variations of intensity within the zone itself, both across and along the beam axis. This zone is followed by a zone where the beam diverges and becomes more uniform (the far field). Fig. 3 illustrates the near field (or Fresnel region) with the transition into the far field (or Fraunhofer region) for cw operation. For a circular piston source of diameter D radiating sound of wavelength λ,

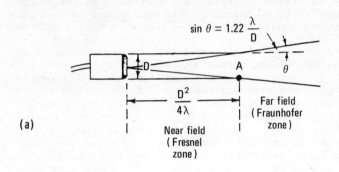

(a)

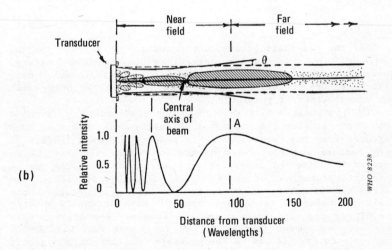

(b)

Fig. 3. Conceptual illustration of the intensity distribution of cw ultrasound: (a) the near- and far-field regions in relation to the transducer, (b) the field distribution of an ideal piston source generating a continuous wave (Adapted from: Wells, 1977).

the Fresnel zone extends from the transducer to a distance equal to $D^2/4\lambda$ (when D is much greater than λ); beyond this distance is the Fraunhofer zone of the transducer. A numerical analysis of the near field of a vibrating piston has been described in the literature (Zemanek, 1971). For a given radius of the transducer, the near field becomes more complex (exhibiting more maxima and minima) as the wavelength of the ultrasound becomes shorter. The acoustic field of a pulsed transducer can be thought of as being composed of contributions from all the frequencies within the bandwidth of a short pulse. It has been shown (Wien & Harder, 1982) that the near field is less structured than that of a cw transducer, and that the length of the near field corresponds to that of a cw transducer oscillating at the centre frequency of the pulsed field.

In the far field of any transducer, the acoustic intensity is proportional to the square of the acoustic pressure. The directivity of the beam in the far field is determined by diffraction, in the same way that a light wave is affected by a small aperture; the higher the frequency of ultrasound produced for a given transducer size, the more directional is the beam. Furthermore, if the frequency is held constant but the diameter is reduced, the beam divergence increases. Equation 2.1 is the formula for conveniently determining the angle of divergence (θ) in the far field (Kinsler & Frey, 1962) as shown in Fig. 3.

$$\text{Sin } \theta = 1.22 \, \lambda/D \qquad\qquad \text{Equation 2.1}$$

For the diagnostic transducers used for pulse echo imaging purposes, the beam width determines the minimum lateral resolution that can be expected. For this reason, many diagnostic transducers are focused to decrease the beam width and enhance lateral resolution.

The intensity distribution along the axis of such a transducer is such that an axial intensity peak occurs at some distance from the transducer. This peak is a common feature of both focused and nonfocused fields, and its existence is an important factor in characterizing ultrasound fields and in the interpretation of some of the biological data. The ultrasonic intensity at this highest main axial peak of the field is referred to as the spatial peak intensity of the field. For exposure in experimental studies, the spatial peak intensity may refer instead to the local maximum, within the exposed region. It is also possible to define a spatial average intensity as the ratio of the power to the beam cross-sectional area, in the plane of interest. The definition of beam cross section (Appendix II) allows a choice of the amplitude at the lateral margin of the beam. Therefore, values of spatial average intensity will depend on this choice and caution should be exercised when comparing reports from different laboratories.

For a theoretical plane circular piston source in an infinite non-reflecting medium, the spatial maximum intensity in the near field is 4 times greater than the spatial average intensity at the transducer surface (Zemanek, 1971; Nyborg, 1977). In actual practice, this ratio typically has values ranging from about 2 to 6 for unfocused transducers, though higher values may be encountered, depending on such factors as the nature of the piezoelectric material used and how it is mounted in the applicator housing (Stewart et al., 1980).

The intensity of the ultrasonic field produced by the transducer also varies with time, if the ultrasound is pulsed. Intensity averaging can be carried out in the time domain and it is therefore necessary to distinguish between time (or "temporal") average (such as the average over the total time or over the pulse duration) and temporal peak intensities (Appendix II).

2.2.2 Standing waves

Standing waves can occur when cw ultrasound is propagating into a confined space, so that the ultrasound waves are reflected back from an interface and travel past each other in opposite directions. This may be the case, for example, within a small room or in a small container of water in the absence of absorbing materials. The resultant waveform, at any instant, is obtained by adding the wave pressures at each point. The acoustic energy distribution is characterized by a stationary spatial pattern with minima and maxima of pressure amplitude, called "nodes" and "antinodes", respectively. Under the conditions applied during medical diagnosis and therapy (generally in the range 1-10 MHz), a progressive wave field usually predominates, though there may be an appreciable standing wave component if, for example, there is a bone/tissue or tissue/gas interface within the beam. The possibility of the occurrence of standing waves is usually of less importance with pulsed ultrasonic irradiation, because they can only exist during the pulse overlap time at a given spatial location.

2.3 Speed of Sound

The speed (c) at which ultrasonic vibrations are transmitted through a medium is inversely proportional to the square root of the product of the density (ρ) and the adiabatic compressibility (B) of the material, such that $c = (\rho B)^{-0.5}$. The speed together with the frequency (f) of the ultrasound determine the wavelength λ ($\lambda = c/f$) of the waves that are propagated. For example, the propagation velocity of ultrasound in most human soft tissues ranges from approximately 1450 to 1660 m/s, so that frequencies of 1 MHz correspond to a wavelength in the range of 1.4-1.7 mm respectively. Thus, ultrasonic diagnostic imaging procedures carried out in this frequency range have the potential for providing resolution of the order of 1 mm. Knowledge of the speed at which ultrasound is transmitted through a medium is used in diagnostic applications for the conversion of echo-return time into the depth of tissue being imaged. Values of sound speed for some other media of interest are given in Table 1 (p. 35) which shows that the speed of sound is highest in solids, somewhat lower in liquids and soft tissues, and very much lower in gases.

2.4 Refraction and Reflection

When an ultrasound wave encounters an interface between two media, the dimensions of which are large compared with the wavelength, part of the wave will be reflected back into the first medium with the same speed. The rest of the wave will be transmitted or refracted into the medium beyond the interface and will travel with the velocity of propagation in that medium (Fig. 4). For reflection, the angles of incidence (θ_i) and reflection (θ_r) are equal; for transmission the angles of incidence and refraction are generally unequal. When the ultrasonic wavelength is equal to or greater than the dimensions of the reflecting object, the incident beam is scattered in all directions.

The ratio of the characteristic impedances (Z_o) of any two media on either side of an interface (see the following section) determines the degree of reflection and refraction or transmission of the incident wave.

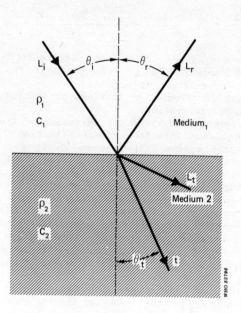

Fig. 4. Reflection (r) and refraction (t) of ultrasound wave incident (i) at the plane boundary between two media 1 and 2. Mode conversion from longitudinal (L_i) to transverse (L_t).

2.5 Characteristic Acoustic Impedance

The characteristic acoustic impedance of a medium is the product of the density (ρ) and the speed ($\underline{c}$) of sound in that medium. The extent to which ultrasonic energy is transmitted or reflected at an interface separating two continuous isotropic media is determined by the ratio of the characteristic acoustic impedances of the media. The closer this impedance ratio is to 1, the more energy is transmitted into the second medium and the less is reflected from the interface. At an interface between human tissue and air, only about 0.01% of the incident energy is transmitted, the remainder being reflected. This illustrates the importance of using a coupling medium between the transducer and human tissue for both therapeutic and diagnostic ultrasound applications. Strong reflections (close to 50%) also occur at bone/tissue interfaces; thus bone/tissue and tissue/gas interfaces constitute an important limitation on the accessibility of some human anatomical regions to diagnostic ultrasonic investigation.

2.6 Attenuation and Absorption

As an ultrasound beam is transmitted through a heterogeneous medium such as soft tissue, its intensity is reduced or attenuated through a number of mechanisms, including beam divergence, scattering, absorption, reflection, diffraction, and refraction.

Beam divergence refers to the spreading of the beam in the far field through diffraction effects (section 2.2.1). For a given transducer radius, this phenomenon is greater at lower frequencies. As the beam area becomes larger, the intensity is reduced.

Scattering refers to the reflection of the incident ultrasound from interfaces (i.e., surfaces separating media of different characteristic acoustic impedances) with dimensions close to or less than the ultrasound wavelength. In this case, the incident beam is scattered in all directions. Ultrasound impinging on blood cells, for example, would be scattered. When scattering occurs, it is greater at higher ultrasonic frequencies.

Absorption of ultrasound occurs when the ordered vibrational energy of the wave is dissipated into internal molecular motion, i.e., into heat. There are many mechanisms by which ultrasound absorption occurs in a medium, including viscous loss, hysteresis loss, and relaxation processes.

The acoustic pressure amplitude $\underline{P}_x$ of the progressive ultrasound wave of initial acoustic pressure amplitude $\underline{P}_o$, at a distance x for a nondiverging beam, in any uniform medium, is described by the relationship:

$$\underline{P}_x = \underline{P}_o e^{-\alpha x} \qquad \text{Equation 2.2}$$

where e is the base of natural logarithms and α is the amplitude attenuation coefficient of the medium (as defined in Appendix I) for a given frequency. α is a measure of the rate at which an ultrasonic wave decreases in amplitude as a function of distance by other than geometric means as it propagates through a medium. For any given medium, α increases with increasing frequency. Because the acoustic intensity is proportional to the square of the acoustic pressure, attenuation can be expressed also in terms of intensity:

$$\underline{I}_x = \underline{I}_o e^{-2\alpha x} \qquad \text{Equation 2.3}$$

Attenuation is important from several points of view. First, it results in a decrease in intensity at various depths in the medium and determines the amount of acoustic energy that can reach structures of interest, either for imaging or therapeutic purposes. Second, attenuation by scattering can

Table 1. Typical values of ultrasonic properties of various media at 1 MHz.

Medium	Ultrasonic speed[a] $\underline{c}$ (m/s)	Characteristic acoustic impedance[b] $\underline{Z}_0 = \rho \cdot c$ (10^3 kg/s m²)	Attenuation coefficient[c] α (Np/cm)	Amplitude absorption coefficient $\alpha_{\underline{a}}$ (Np/cm)
air (dry) (20 °C)	343.6	0.45	0.18	0.18
water (37 °C)	1480	1480	0.0002	0.0002
amniotic fluid	1530–1540	1540–1560	0.0008	ND
aqueous humour) vitreous humour)	1530–1540	1540–1560	0.005–0.08	ND
blood) plasma)	1555–1525	1560–1580	0.001–0.002	ND
testis			0.03–0.04	0.01–0.02
fat	1450–1490	1360–1400	0.07–0.24	ND
liver) kidney) brain) heart)	1560–1600	1580–1620	0.07–0.3	0.02–0.05
spleen) pancreas)	1510–1600	1580–1620	0.07–0.3	ND
muscle	1560–1600	1620–1700	0.06–0.16	ND
uterus	1600–1660		0.02–0.20	ND
lens	1600–1660		0.02–0.20	ND
skin) tendon)	1720–2000		0.04–0.50	ND
bone	3000–3300	4000–7000	1.3–3	ND
lung	500–1000		2–3	ND

Note: These values are for animal tissue and are for illustrative purposes only;
    published data are not always consistent. Actual measured values may show
    quite strong variability with factors such as tissue preparation
    temperature and intensity.
[a] Velocity of longitudinal waves.
[b] Estimated from published data.
[c] Attenuation is approximately proportional to frequency: $\alpha = \alpha_1 \underline{f}^m$,
  where α_1 is the attenuation coefficient at 1MHz, $\underline{f}$ is the frequency in
  MHz, and known values of m lie between 0.76 (tendon) and 1.14 (brain).
ND = not determined.

result in ultrasonic energy reaching unintended structures. Third, attenuation is important, because it is due in part to an absorption process in which the propagating energy is permanently modified (for example, converted into heat energy which causes a temperature rise in tissue). In therapeutic applications, energy absorption and heat generation in tissue are usually the intended results.

Attenuation is greater in some soft tissues than in others. This variation is exploited in therapy for differential absorption and heating of ligaments and tendons in surrounding muscular tissue (Lehmann et al., 1959; Stewart et al., 1982).

Because of the depth of penetration desired, the frequencies used for therapy purposes range from about 0.5 to 3 MHz. For diagnostic purposes, the upper limit of the range for imaging in abdominal areas is about 10 MHz. Frequencies up to 20 MHz are used for small structures such as the eye, which have a lower attenuation coefficient and shorter imaging depth.

Absorption is considerably higher in bone than in soft tissues. This is one reason why bone may constitute a critical organ for some forms of ultrasonic exposure, especially ultrasound therapy, even though there is a strong reflection from a bone/soft tissue interface. Bone damage has been reported in experimental animals (Barth & Wachsmann, 1949; Kolar et al., 1965) at levels just higher than those normally employed in physiotherapy (i.e., 3-4 W/cm^2) (section 6.4.6). In addition, ultrasound exposure of a bone/tissue interface can result in sudden and sometimes pronounced periosteal pain arising from a buildup of heat at the interface. At the bone/tissue interface, some of the longitudinal oscillations (particles oscillating in the direction of propagation) are transformed into transverse oscillations. The transverse oscillations (shear waves) are more readily absorbed than longitudinal waves. This can produce local heating at the bone/tissue interface causing periosteal pain (Lehmann et al., 1967).

2.7 Finite Amplitude Effects

Another effect that may be important when ultrasound is applied in biomedical research, diagnosis, or surgery results from the finite amplitude of the particle velocity of the ultrasonic wavefront. In linear· acoustics, two familiar assumptions are made: (a) that the transmitted frequency is the only frequency produced; and (b) that when the input amplitude is increased, the amplitude at remote points in the field increases proportionally. These linear assumptions are not valid when considering finite-amplitude effects. For a more detailed explanation, the reader is referred to Beyer & Letcher (1969).

It has been shown (Beyer & Letcher, 1969; Muir & Carstensen, 1980; Carstensen et al., 1981) that the frequencies and intensities used in pulsed diagnostic ultrasonics can potentially create significant distortion of sound waves in water.

3. MECHANISMS OF INTERACTION

When acoustic energy is absorbed by matter, it is converted into heat, the consequent temperature elevation depending on the amount of energy absorbed, the specific heat of the medium, and the dynamic balance between heat deposition and removal. In contrast to X-rays, for example, commonly used ultrasound beams can carry appreciable amounts of energy and thus one mechanism of action of potential biological importance is thermal. A second phenomenon that is well known to be associated with ultrasonic energy, and to play a major role in many of the biological changes that have been induced by ultrasound applied in vitro, is cavitation. However, not all the evidence of biological and biochemical changes induced by ultrasound can be explained on the basis of either heat or cavitation. It is necessary to be aware of a further group of established and/or physically predictable stress mechanisms, and of the possible existence of other biophysical mechanisms, hitherto undocumented. Finally, it should be noted that the different mechanisms, as classified in this manner, are not necessarily independent; for example, the biological expression of a physical stress directly induced by the passage of ultrasound may well be influenced by the temperature of the irradiated structure. Examples of reviews of ultrasound mechanisms are those published by Nyborg (1977, 1979, 1982) and Repacholi (1981).

3.1 Thermal Mechanism

Several reviews concerning the elevation of temperature resulting from ultrasound exposure have been published (Lele, 1975; Nyborg, 1977).

When ultrasound interacts with matter, part of the energy of the beam will be absorbed and converted into heat. The rate (Q) at which heat is generated per unit volume within a medium is given by the equation $Q = 2I_a\alpha_a$; where α_a is the amplitude absorption coefficient of the medium and I_a is the intensity of a plane travelling ultrasound wave (Appendix I). Without heat conduction away from the exposed site, the rate of temperature rise will be (Dunn, 1965):

$$dT/dt = 2\alpha_a I_a / \rho C_m \qquad\qquad \text{Equation 3.1}$$

where $d\underline{T}/d\underline{t}$ is the temperature rise per unit time, ρ is the ambient density of the medium, and $\underline{C}_m$ is the specific heat per unit mass.

Consider an example of soft tissue exposed to an ultrasound beam of intensity 1 W/cm^2. If $\rho = 1$ g/cm^3, $\underline{C}_m = 1$ $cal/g/^\circ C$ and α_a is 0.1 Np/cm, the temperature rise $d\underline{T}/d\underline{t}$ is then 0.048 $^\circ C/s$, when heat conduction is neglected.

If the effect of heat conduction away from exposed matter is considered, it will be appreciated that, following an initial rise, the temperature will tend towards an equilibrium value. Calculations covering this behaviour for a spherical model have been given by Nyborg (1977); some results are shown in Fig. 5. For this model (a spherically symmetrical object exposed in an isotropically conducting medium), the increase in equilibrium temperature is proportional to the square of the radius, as is the time required to attain that temperature. Thus, a small body uniformly exposed to ultrasound will experience a small but rapid temperature rise, whereas a large body, uniformly exposed to the same ultrasound

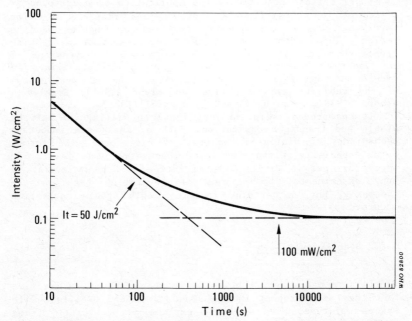

Fig. 5. Computed "thermal threshold" curve for a spherical absorber with an absorption coefficient = 0.1 Np/cm^2 and a radius of 1.2 cm; conductivity coefficient as for water. The curve shows the intensity ($\underline{I}$) required to produce a temperature increase of 2.4 $^\circ$C at the centre of a sphere after various exposure times ($\underline{t}$) (From: Nyborg, 1977).

intensity, will reach a higher final temperature, but over a longer period of time. It follows that temperature elevations resulting from local heating on a scale comparable to cellular dimensions (10-50 μm), which presumably occurs as a result of local absorption mechanisms, will be insignificant in practice. This conclusion was reached independently by Love & Kremkau (1980).

In practice, the biological expression of heat-induced damage is found to depend both on the maximum temperature achieved and on the time period for which that temperature is maintained. According to Lele (1975), exposure of mice to a temperature elevation of 2.5-5.0 °C for an hour or more during pregnancy caused a significant increase in the number of fetal abnormalities.

3.2 Cavitation

3.2.1. Introduction

Under certain conditions, the application of ultrasound to a liquid (or quasi-liquid) medium gives rise to activity involving gaseous or vaporous cavities or bubbles in the medium. This phenomenon, termed cavitation, may require pre-existing nuclei, i.e., bodies of gas with dimensions of the order of micrometres or smaller which are stabilized in crevices or pores, or by other means, in the medium. Reviews of the subject have been given by Flynn (1964), Coakley & Nyborg (1978), Neppiras (1980), and Apfel (1981).

It has proved useful (Flynn, 1964) to distinguish between stable and transient cavitation. Both of these are important mechanisms for biological effects of ultrasound, the former being especially relevant at lower intensity levels (e.g., 300 mW/cm² or less in water) and the latter at higher levels. In many experiments, both types of cavitation occur simultaneously, but in certain situations only stable cavitation occurs.

3.2.2 Stable cavitation

In some media, gas bubbles exist which are of such a size that they are resonant in the sound field and oscillate with large amplitude. When a bubble expands and contracts during the ultrasound pressure cycle, the surrounding medium flows inwards and outwards with a higher velocity than if the gas bubble were absent. As a rough guide, the resonant diameter of

a cavitation bubble in water at 1 MHz is about 3.5 µm.
Alternatively, gaseous nuclei may exist in the medium which
are initially smaller than resonance size but which grow to
that size in an applied sound field through the process of
rectified diffusion.

When a gas bubble pulsates, its motion is not usually
spherical, either because of distortion by an adjoining
boundary or because of surface waves set up by the ultrasound
field. Asymmetric or non-uniform oscillation of the air-liquid
interface, at the surface of an air pocket or bubble, causes a
steady eddying motion to be generated in the immediately
adjoining liquid, often called microstreaming, in which the
velocity gradients may be high. If biopolymer molecules or
small biological cells are suspended in liquid near a
pulsating bubble, they may be swept into a region of high
velocity gradient. Such a situation can also occur if a small
bubble pulsates near a cell membrane causing the membrane to
vibrate, producing streaming motions within the cell. The
biological system will then be subjected to shearing action
and damage may occur, such as fragmentation of macromolecules
and membranes (Nyborg, 1977).

Significant biological effects occur in suspensions near
resonant bubbles, even at low spatial peak temporal average
(SPTA) intensity levels. For example, Barnett (1979), and
Miller et al. (1979) found that blood platelets tended to
aggregate around artificial holes (forming gas bubbles) in a
membrane, and Williams & Miller (1980), using similar membrane
material (containing gas-filled pores) observed ATP release
from red blood cells. All of these effects were observed at
SPTA levels considerably lower than 0.1 W/cm^2.

These findings are consistent with the theory of
microstreaming and with experimental information on the
response of biological cells to hydrodynamically generated
viscous stress (Glover et al., 1974; Brown et al., 1975;
Anderson et al., 1978; Dewitz et al., 1978, 1979). For example
Nyborg (1977) estimated that a bubble of 3 µm radius in
blood plasma, caused to pulsate by ultrasound at an intensity
of 1 mW/cm^2 with a frequency of about 1 MHz (to which the
bubble is resonant), would generate a microstreaming field in
which the maximum viscous stress would greatly exceed 100
N/m^2. The latter is an intermediate value for
hydrodynamically generated viscous stress which causes cell
lysis.

Pulsating bubbles also produce microstreaming in organized
tissues. Martin et al. (1978) reported acoustic streaming
motions in plant and mammalian systems, using Doppler fetal
heart monitors under experimental conditions that ensured the

existence of gas bubbles. According to Akopyan & Sarvazyan (1979), streaming can produce changes in the relative positions of intracellular organelles and breaks in cytoplasmic structures.

3.2.3 Transient cavitation and studies concerned with both stable and transient cavitation

In contrast to stable cavitation, transient (or collapse) cavitation is more violent and occurs at higher ultrasound intensity levels. When a gas bubble or nucleus within the medium is acted on by an ultrasound field having a high pressure amplitude, it may expand to a radius of twice the original value or more, then collapse violently. In the final stages of collapse, kinetic energy given to a relatively large volume of liquid has to be dissipated in an extremely small volume, and high temperatures and pressures result. Idealized thermodynamic calculations show that for a compression in which no heat escapes from the cavity at the end of the cavity's existence, the final temperature is around 8000 K and the pressures are greater than 10^9 Pa (10^4 atmospheres). Of course, the idealized assumption of a thermodynamically closed system is not valid under such extreme conditions. Sutherland & Verrall (1978) report that, under actual conditions, not all the heat remains trapped in the cavity during collapse; some is conducted away, resulting in estimated temperatures of the order of 3500 K. It seems reasonable to assume that effects on biological systems may be induced at least by the mechanical shock waves and high temperatures generated during the bubble collapse.

Chemical changes are commonly produced by cavitation. The combination of high pressures and temperatures can generate aqueous free radicals and hydrated electrons (highly reactive chemical species) within the exposed medium by the dissociation of water vapour in the bubble during its contraction. Chemical interactions of biomacromolecules with these free radicals often result (especially with hydrogen $H^{\cdot}$ and hydroxyl $OH^{\cdot}$ radicals), and significantly alter their properties. This can be accompanied by the formation of such compounds as nitrous acid (HNO_2), nitric acid (HNO_3), and hydrogen peroxide (H_2O_2) (Akopyan & Sarvazyan, 1979).

Studies show that transient cavitation does not occur unless the intensity exceeds some threshold value which is very dependent on experimental conditions. The cavitational threshold SPTA intensity was determined by Esche (1952) and Hill (1972a) for frequencies ranging from 0.25 to 4 MHz, in air-equilibrated water, for cw ultrasound. The threshold intensity was in the range of a few watts per square centimetre and was frequency dependent. The higher the frequency, the higher the intensity required to produce cavitation.

Pulsing conditions have a marked influence on cavitation. Hill & Joshi (1970) found that, at shorter pulse durations, the cavitation threshold increased. Alternatively, as the pulse duration decreased, the duty factor had to be increased to produce cavitation at a given intensity. A model for acoustic cavitation, according to which cavitation activity is optimized for an appropriate choice of pulsing parameters, has been postulated and confirmed experimentally by Ciaravino et al. (1981).

Higher ambient pressure causes higher threshold intensities for cavitation. For a cw 1 MHz ultrasound beam, Hill (1972a) found that the threshold intensity varied from just under 1 W/cm^2 at an ambient pressure of 10^5Pa (1 bar) to much greater than 16 W/cm^2 at 1.75×10^5Pa (1.75 bar). Increasing the ambient pressure often provides an effective means of inhibiting cavitation and thereby ascertaining whether a previously observed response was due to cavitation.

It has also been found that the threshold for cavitation decreases with increasing temperature (Connolly & Fox, 1954) and with increasing volume of the irradiated liquid (Iernetti, 1971).

Particularly important for the occurrence of cavitation is the number and size distribution of gas nuclei within the medium. Unfortunately, these quantities are not easily measured. The number of available nuclei within a fluid medium greatly increases when the medium is stirred or mechanically disturbed (Williams, 1982a).

3.2.4 Cavitation in tissues

Intracellular gas channels are commonly present in plant tissues and greatly influence the biological response of these tissues to ultrasound (Nyborg et al., 1975; Carstensen, 1982). Similarly, the responses of insects and insect eggs to ultrasound are greatly influenced by the presence of microscopic airpores (Child et al., 1980a, 1981a, 1981b). A characteristic of the response of both plants and insects to pulsed ultrasound is that the critical exposure parameter appears to be the temporal peak rather than the temporal average of the intensity.

Much less is known about cavitation in mammalian tissues. In a series of studies, Fishman (1968) was unable to detect significant levels of haemolysis in the blood of human volunteers whose hands were immersed in an 80 kHz cleaning bath for up to 45 min. However, the external ears of rabbits developed numerous petechial haemmorrhages when they were immersed for more than 3 min in a 55 kHz cleaning bath (Carson & Fishman, 1976).

Lehmann (1965a), using dogs, reported that tissue damage, which was attributed to cavitation, occurred at intensity thresholds of 1-2 W/cm^2 for 1 MHz ultrasound applied by means of a stationary applicator. When a stroking technique was used, these effects were not observed at intensities up to 4 W/cm^2. A dependence on ambient pressure, observed for this biological effect is a strong indication that the gas content of the tissue was involved in the reaction. Thresholds of about 1.5 W/cm^2 have been reported for soft tissue damage due to cavitation caused by exposure to cw ultrasound with the transducer in a stationary position (Hug & Pape, 1954). On the basis of morphological findings and physical measurements, they concluded that cavitation could be expected in tissues at intensities in the range used for therapeutic purposes. Similar data have also been reported by Lehmann & Herrick (1953). Other reports of effects on experimental animals also indicate that cavitation may have been responsible (O'Brien et al., 1979; Martin et al., 1981).

Evidence for the existence of gaseous nuclei in tissues has been given by ter Haar & Daniels (1981). They observed that the production of gas bubbles in the legs of guinea-pigs exposed to cw 0.75 MHz ultrasound at SATA intensities of 80 and 680 mW/cm^2, was associated with tissue interfaces. At 680 mW/cm^2, sites occurred throughout the entire cross-section of the leg with many bubbles located intramuscularly. The rate of appearance of sites increased with both intensity and duration of exposure. The authors reported that an SATA intensity of 80 mW/cm^2 appeared to be close to an intensity threshold for stable bubble production in tissues in vivo. In applying the theory for rectified diffusion to these results, Crum & Hansen (1982) showed that they were consistent with an assumption that gaseous nuclei with diameters in the range of a few micrometres exist normally within tissues.

3.3 Stress Mechanisms

Stress mechanisms or non-thermal, non-cavitational mechanisms of ultrasound action have been reviewed by Nyborg (1977) and Dunn & Pond (1978). Ultrasound exposure produces various stresses within biological systems, the magnitude and significance of which depend on the detailed characteristics of the ultrasound field and the biological system exposed. Lewin & Chivers (1980) proposed a viscoelastic model of the cell membrane as a potential means of investigation in connection with pulsed sources. Repacholi (1982) found evidence that many biological effects on cell systems in vitro may be due to forces both within and outside the cell, which might be mediated by stress mechanisms.

Stresses or forces resulting from an ultrasound field acting on heterogeneous regions within a medium can be categorized as follows (Dunn & Pond, 1978):

(a) buoyancy forces that are oscillatory, have a time-average equal to zero, and produce a radiation pressure on bodies having a density different from the surrounding medium;

(b) displacement or radiation forces that have a non-zero time average and can cause an appreciable relative velocity between the inhomogeneity and the surrounding medium;

(c) viscosity-variation forces or viscous stresses that result in acoustic streaming because of variations in viscosity over the cycle of the applied ultrasound; and

(d) the Oseen force, another time-averaged force, which is due to the dependence of drag on the second power of relative velocity.

3.3.1 Radiation pressure, radiation force, and radiation torque

There is evidence of radiation pressure (from ultrasound pulses) being detected by the inner ear and giving rise to disturbances that can be sensed by the brain as if they were audible sound (Foster & Wiederhold, 1978). In addition, Gershoy & Nyborg (1973) postulated that gradients of radiation pressure in exposed plant tissue give rise to water flow in cytoplasmic channels.

An example of the action of radiation force is the blood flow stasis phenomenon reported by Dyson et al. (1971), where red blood cells in the blood vessels of chick embryos exposed to an ultrasonic standing-wave field, collected into parallel bands spaced at half wavelength intervals. This has also been shown in mammalian vessels (ter Haar et al., 1979).

Spinning of intracellular bodies exposed to highly non-uniform ultrasound fields has been observed by various investigators (Dyer, 1965, 1972; Nyborg, 1977; Martin et al., 1978). When an ultrasound field is propagated within a liquid, a twisting action may be exerted on suspended objects, and on elements of the liquid itself. For an asymmetrically shaped object such as a rod or disc, this radiation torque varies

with the orientation of the object relative to the oscillation
direction of the surrounding liquid, so that the object tends
to assume the position in which the torque on the object is
least. Such an effect may be important, when the effects of
ultrasound on asymmetrically shaped cells, organelles, or
macromolecules are considered. For a symmetrical object,
steady spinning will result. Theoretically, this spinning is
expected in non-uniform fields such as those existing at a
boundary where a progressive ultrasound wave impinges
obliquely and is reflected (Nyborg, 1977). In the latter
situation, the object's velocity of spinning ($\underline{v}$) is
proportional to the ratio of the absorption coefficient
(α_a) for the material in this spherical body and to the
coefficient of shear viscosity (η) for the surrounding fluid.

Martin et al. (1978) observed the effects of radiation
torque in sonicated (2.1 MHz, 43 mW/cm^2) leaves of Elodea
and root tips of Vicia faba. How radiation torque affects
other macromolecular structures or organelles within or
outside cells is not known, at present.

3.3.2 Acoustic streaming

When an ultrasound field is propagated within a liquid,
the particles of the liquid take part in an oscillatory flow.
Consider a particle oscillating in a direction parallel to a
boundary. At the boundary itself, the velocity of the liquid
flow will be zero provided the boundary is a fixed, rigid
solid, and "non-slip" conditions apply. Conditions may then
exist for establishing acoustic streaming, a time-independent
circulatory motion of the liquid. As part of this motion a
thin boundary layer may exist between the surface and the
streaming liquid itself, within which the velocity gradient is
large. Such streaming has been observed as circulatory flow in
the vacuoles of plant cells (Nyborg, 1978). However, there
must be non-uniformity or some kind of asymmetry for this
streaming to be established. For an ultrasound field
propagating in a suspension of particles, relative motion
occurs between the particles and the fluid, where boundary
layers are established around each particle and give rise to
an acoustic streaming field. Such microstreaming was
demonstrated near vibrating gas bubbles by Elder (1959), who
analysed four regimes of streaming.

Early effects attributed to acoustic streaming were
reported by Nyborg & Dyer (1960), who demonstrated the
migration of protoplasm towards a needle vibrating at 25 kHz
in intact cells of Elodea. Selman & Jurand (1964) described
the disorganization and subsequent recovery of the arrangement

of the endoplasmic reticulum following irradiation for 5 min with 1 MHz ultrasound at intensities between 8 and 15 W/cm². More recently, these stresses associated with acoustic streaming have been suggested to be responsible for:

(a) altered cell surface charge (Repacholi, 1970; Repacholi et al., 1971; Taylor & Newman, 1972);

(b) altered cell membrane permeability (Chapman, 1974; Chapman et al., 1980; Al-Hashimi & Chapman, 1981);

(c) separation of small fragments from cells (Dyson et al., 1974; Nyborg, 1979; ter Haar et al., 1979);

(d) rupture and fragmentation of cell membranes (Williams, 1971; Brown et al., 1975; ter Haar et al., 1979); and

(e) reduced uptake of radioactive precursor in mammalian cells in vitro (Repacholi, 1980).

4. MEASUREMENT OF ULTRASOUND FIELDS

The spatial distribution of ultrasound fields can be quite complicated depending on such factors as focusing, the radius of the transducer, the wavelength of the ultrasound, the distance from the source, and even on the way in which the element of the transducer is mounted (Zemanek, 1971). Any effect produced by ultrasound will depend quantitatively on the temporal and spatial characteristics of the ultrasonic field. It is therefore necessary to consider the methods available for making physical measurements to determine the relationships between the equipment output levels used in human exposure and the results of biological studies.

These methods are divided into measurement techniques for liquid-borne and airborne ultrasound. Several extensive reviews of techniques for measuring liquid-borne ultrasound have been reported in the literature (Stewart, 1975, 1982; Zieniuk & Chivers, 1976). The phenomenon of solid-borne ultrasound, for example, in bone (Fry & Barger, 1978) is also of interest, but will not be dealt with here.

4.1 Measurement of Liquid-borne Ultrasound Fields

Measurements necessary to characterize ultrasound fields should include all spatial and temporal characteristics. This will involve measuring at least one (and possibly more) of the four field parameters ($\underline{p}$, ξ, $\underline{v}$, $\underline{a}$), discussed in section 2, over all relevant conditions of space and time. Once these parameters are known, it is possible to calculate the spatial and temporal behaviour of power and intensity in the equivalent plane-wave field. In order to characterize exposure, the total power should be specified as well as the following intensities: spatial average temporal average (SATA) intensity; spatial peak temporal peak (SPTP) intensity; spatial peak temporal average (SPTA) intensity; and, if applicable, spatial peak pulse average (SPPA) intensity and spatial average pulse average (SAPA) intensity. These and other factors that are important for the complete characterization of ultrasonic exposure in the investigation of biological effects are summarized in Table 2.

Acoustic power and intensity have traditionally been used to express exposure. They are the parameters specified in most standards, e.g., the AIUM-NEMA (1981) standard, the Japanese standards for diagnostic equipment (JIS 1979, 1980, 1981; JAS, 1976, 1978), and the standards of Canada (Canada, Department of

National Health and Welfare, 1981) and the USA (US Food and Drug Administration, 1978) for the performance of ultrasound therapy equipment.

Table 2. Biologically important exposure parameters

(a) Continuous wave (cw) ultrasound

Frequency of ultrasound
SATA intensity
SPTA intensity (if focused)

(b) Pulsed ultrasound

Centre frequency
Pulse shape or frequency spectrum
Pulse duration
Pulse repetition frequency or duty factor
Frame repetition frequency (automatic scanners)

SPTP intensity
SPPA intensity
SPTA intensity

(c) General

Exposure time
Exposure fractionation (if not a single exposure)
Degree and periodicity of the modulation or interruption
Single transducer
Transducer diameter
Array dimensions (automatic scanners)
Type of field (focused or unfocused)
Focal area, focal length (if focused)
Other details of geometric conditions, such as:
Exposure under far-field or near-field conditions
Acoustic path length to organ or site of interest
Extent of standing wave component (if any)
Relation of the peak to the average intensity for
 the beam cross section of interest, (i) if the source is
 maintained in a fixed position and orientation during exposure;
 (ii) if not fixed, the path and speed of motion

Relatively little work has been carried out concerning ultrasonic field measurements in tissue, though some measurements and theoretical calculations to determine the ultrasonic field in tissue have been reported (Chan et al., 1974). Instrumentation used for internal field measurements include thermocouples for the measurement of temperature rise at specific locations (Goss et al., 1977) and miniature transducers inserted into bodies (Bang, 1972; Lewin, 1978).

4

Reported measurements of the attenuation between the abdominal surface and the uterine cavity are shown in Table 3.

Table 3. Reported attenuation between the abdominal surface and the uterine cavity[a]

No. patients	Average rate of attenuation (dB/cm)	Attenuation (dB)	Distance (cm)	Frequency (MHz)	Species	Reference
10		1.6 (mean)		2.25	mouse	Bang & Northeved (1970)
8	0.5 - 1	2 - 4	2 - 4.5	2.25	man	Bang (1972)
6	0.9 - 1.56	6 - 14	5 - 11	2.25	man	Etienne et al. (1976)
13	0.6 - 1.8	2 - 7.5	3 - 5.8	2.25	man	Takeuchi et al. (1977)
10	0.5 - 7.2	12 (mean)	6	2.0	man	Morohashi & Iizuka (1977)

[a] From: Stewart & Stratmeyer (1982).

Instruments available for measuring liquid-borne ultrasound include those that measure total power and those that can measure point quantities over an area. With the latter, it is possible to determine the distribution of the energy in the ultrasonic field.

4.1.1 Measurement of the total power of an ultrasound beam

Measurement of total power is important for several reasons: (a) the total power of an ultrasound field impinging on an extended plane target can generally be measured more accurately than point or spatial quantities; (b) it is commonly used to characterize standard reference sources (such sources may be used in the calibration of detectors that measure point quantities, e.g., hydrophones); and (c) on measuring the total power for a defined field size, it is possible to calculate the mean intensity, usually referred to as spatial average intensity.

Ultrasound measurement procedures are discussed by various authors (O'Brien, 1978; Stewart, 1982). Several methods are available for measuring total power, including radiation force, calorimetry, and acoustico-optical techniques, but the one which is usually favoured is radiation force. This method, which can be used in the measurement of the total power output of ultrasound equipment, is based on the fact that the surface of a reflecting or absorbing target is performing a micro-scopic oscillation according to the continuity of particle velocity ($\underline{v}$) and partitioning of the momentum carried by the plane wave takes place at the surface. Consequently, the time average of the acoustic pressure at this non-stationary reference surface is non-zero. The resulting steady pressure on the surface, multiplied by the exposure area, is called the radiation force. The force produced is independent of frequency and is proportional to the total ultrasonic power impinging on the target. The radiation force ($\underline{F}$) in newtons is given by:

$$\underline{F} = \underline{P}D/\underline{c} \qquad\qquad \text{Equation 4.1}$$

where $\underline{P}$ is the incident acoustic power in watts, $\underline{c}$ is the propagation velocity of the wave in m/s (in water $\underline{c} = 1.5 \times 10^3$ m/s at 30 °C), and D is a dimensionless factor, determined by the type of interface encountered by the ultrasonic field and the direction in which the force produced by reflection or absorption is measured.

Values for D in Equation 4.1 are shown in Table 4. The table has been modified from that of Hueter & Bolt (1955) to a more general situation (Stewart & Stratmeyer, 1982). By knowing the type of interface a target presents to an ultrasonic field, and by measuring the magnitude of the force the total power in the acoustic field can be computed. Typically, a flat, totally reflecting plate is used in radiation force devices. For this situation, the only force produced by the reflected ultrasound is in a direction normal to the plate. This force is given by $2\underline{P}/\underline{c} \cos \theta$, where θ is the angle between the normal to the reflecting surface and the ultrasound beam. If the direction of measurement of force is not normal to the plate, only the component in the direction of measurement will be determined. In this case, the force measured is $F = 2\underline{P}/\underline{c} \cos \theta \cos \psi$, where ψ is the angle between the normal to the reflecting surface and the direction in which the force is to be measured.

If $\theta = \psi$, i.e., the ultrasound beam and the direction in which the force is measured are the same, then $F = 2\underline{P}/\underline{c} \cos^2\theta$, which is the equation usually associated with these devices (Hueter & Bolt, 1955). For propagation in water, a collimated beam of ultrasound exerts an apparent weight in the direction of propagation equivalent to $0.136 \cos^2\theta$ mg/mW or 0.067 mg/mW for $\theta = 45°$.

Table 4. Values of the constant D for various physical situations for a plane progressive ultrasound field[a]

Physical situation	$D^{\underline{x}}$	$D^{\underline{y}}$
Perfect absorber, normal incidence[b] $r = 1$	1	$1 \cos \psi$
Perfect reflector, normal incidence $r = 0$ or ∞	2	$2 \cos \psi$
Perfect reflector, ultrasound incident at angle θ to reflector[b] $r = 0$ or ∞	$2 \cos^2 \theta$	$2 \cos \theta \cos \psi$
Nonreflecting interface, normal incidence[b] $r = 1$, $c_1 \neq c_2$	$1 - c_1/c_2$ For $c_1 < c_2$, force in direction of propagation For $c_1 > c_2$, force directed opposite to direction of propagation	$(1 - c_1/c_2) \cos \psi$
Partially reflecting interface, normal incidence $Z_2 \neq Z_1$, $c_1 \neq c_2$	$2[(r-1)^2/(r+1)^2]$	$2[(r-1)^2/(r+1)^2] \cos \psi$

[a] From: Hueter & Bolt (1955) and Stewart & Stratmeyer (1982).
[b] $r = Z_2/Z_1$, the impedance ratio at an interface, where $Z = \rho c$.
[x] where the direction of ultrasound propagations is the same as the direction in which the force is measured.
[y] where the direction of ultrasound is not the same direction in which the force is measured.
[c] = the velocity of ultrasound in the medium.
ρ = the density of the medium.
θ = the angle between the normal to the reflecting surface and the incident ultrasound beam axis.
ψ = the angle between the normal to the reflecting surface and the direction in which the force is measured.

Note:
(1) When the direction of the incident ultrasound beam is the same as the direction in which the force is measured, then $\psi = \theta$ and the value of D for a reflecting surface becomes $2 \cos^2 \theta$; this is usually the case in practice.
(2) When the direction in which the force is measured is the same as the direction of the normal to the reflecting surface, then $\psi = 0$ and the value of D for a reflecting surface becomes $2 \cos \theta$.

The relationship in equation 4.1 applies for both cw and pulsed ultrasonic fields, provided that $\underline{P}$ is taken as a time-averaged value. Because of inertia, the system cannot respond to the temporal variation of the pulsed ultrasound, unless the pulse repetition rate is extremely slow. Many practical radiation force systems for measuring the output from both therapy and diagnostic sources have been described in the literature (Rooney, 1973; Stewart, 1975; Robinson, 1977; Brendel et al., 1978; Carson et al., 1978; Bindal & Kumar, 1979, 1980; Bindal et al., 1980; Carson, 1980; Shotton, 1980).

4.1.2 Spatial and temporal measurements

Ideally, to measure the spatial and temporal characteristics of ultrasound, a detector is needed that is small compared with the wavelength of the ultrasound field and has a response function (i.e., the quotient of the electric output signal and the acoustic imput signal) that is flat over the frequency of interest, combined with high sensitivity, low noise, and a wide acceptance angle. Miniature piezoelectric hydrophones, though not ideal, are used extensively to determine the spatial distributions and temporal pressure waveforms and, when properly calibrated against an appropriate standard, can provide a satisfactory measurement method. Wells (1977) describes various types of hydrophones that have been used. Devices of this type respond to the instantaneous local value of the acoustic pressure in the field. However, not all commercially available hydrophones are frequency independent in their sensitivity, and this presents a major problem. The frequency responses of several hydrophones have been reported in the literature (Harris et al., 1977; Lewin, 1978, 1981a, b; Harris, 1981).

The International Electrotechnical Commission (IEC, 1981) and the American Institute for Ultrasound in Medicine/National Electrical Manufacturers Association joint task group (AIUM-NEMA, 1981) have both recommended the use of hydrophones for the measurement of spatial and temporal exposure parameters for diagnostic ultrasound equipment. Comparison of the reciprocity technique for the calibration of ultrasonic hydrophones with that of planar scanning in a field of known acoustic power has shown that both methods yield consistent results (Gloerson et al., 1982). The choice of method depends on convenience and the interest and background of the user.

Most conventional probes have resonances in the frequency range of interest but distort the ultrasonic pulses being observed. Only if the frequency characteristics of the probe

are known, can appropriate corrections be made. Another
limitation in the use of hydrophones is their directional
sensitivity, for which correction must be made. The use of the
piezoelectric polymer polyvinylidene fluoride as an ultrasonic
hydrophone has been described (DeReggi et al., 1978, 1981;
Wilson et al., 1979; Shotton et al., 1980; Harris, 1981;
Lewin, 1981b). Compared with ceramic, this material has an
acoustic impedance much closer to that of water and, because
it is available in sheets that have thickness resonances
greater than 20 MHz, it promises to be useful as a broad-band,
acoustically transparent receiver. Hydrophones made with
piezoelectric polymer are commercially available.

4.2 Measurement of Airborne Ultrasound Fields

Both audible and ultrasonic fields are usually quantified
in terms of sound pressure level (SPL), in decibels (dB):

$$SPL \ (dB) = 20 \ \log_{10}(\underline{p}/\underline{p}_r)$$

where p is the acoustic pressure in free air. The reference
pressure $\underline{p}_r$ is usually taken as $\underline{p}_r = 20$ micropascals
(μPa), which is equivalent to an acoustic intensity of
$\underline{I}_r = 10^{-12} W/m^2$. This is approximately the lowest intensity
of audible sound perceived by human subjects at 1000 Hz.

Since acoustic intensity is proportional to the square of
acoustic pressure, the sound level can equally be expressed by:

$$SPL \ (dB) = 10 \ \log \ (\underline{I}/\underline{I}_r)$$

Therefore, doubling the intensity $\underline{I}$ increases the SPL by 3
dB, whereas doubling the pressure p increases the SPL by 6 dB.

The actual determination of decibel levels at various
positions in an airborne ultrasound field can be made with
several commercially available systems (Michael et al., 1974;
Herman & Powell, 1981). These normally include a capacitor
microphone sensing element having a flat frequency response
within the range of interest, and signal processing circuitry.
Usually, this circuitry includes a set of one-third octave
filters, so that the additive SPL within any particular
one-third octave frequency range is indicated on the meter. A
spectrum of SPL as a function of frequency (to one-third
octave resolution) can be obtained by "stepping through" the
filter set. When making SPL measurements, humidity and
temperature conditions should be taken into account.

Rapid advances are being made in the development of ultrasound transducers for use in air, which have greatly improved resonance frequency and resolving capacity. Commercially available transducers include electrostatic types, with linear frequency ranges up to a few hundred kHz (Frederiksen, 1977) and ceramic types, with quarter-wavelength matching to air and resonant frequencies up to 400 kHz (Kleinschmidt & Magori, 1981). At these frequencies, the ultrasound wavelength in air is of the order of 1 mm, which enables the construction of a whole line of new instrument systems using very narrow ultrasound beams (mm to cm) for remote measurements over distances ranging from millimetres to metres.

Applications using measurement of airborne ultrasound include: industrial remote measurements (size, location, speed etc.), anthropometrical measurements, and imaging of human beings (Lindström et al., 1982). Measurements are performed using the ultrasound pulse-echo method, which means that many techniques used in diagnostic ultrasound can be transferred to high-frequency airborne ultrasound, i.e., different forms of real-time scanners (Lindström & Svedman, 1981).

Systems developed for measurement, control and imaging, and working with high-frequency (50-1000 kHz) airborne pulse-echo ultrasound, make use of narrow sound beams of high pulse intensity but low duty rate (Lindström et al., 1982). Because of the short pulse duration, determination of the intensity level should be performed in a similar way to the procedure for diagnostic ultrasound; i.e., using spatial and temporal measurements to characterize the airborne ultrasound field.

5. SOURCES AND APPLICATIONS OF ULTRASOUND

For many years, ultrasound was only used in the detection of submarines (Mason, 1976). The device, first produced by Paul Langevin in 1917, was composed of a quartz crystal vibrating at 50 kHz, propagating ultrasound into the water and detecting the reflected beam. Ultrasound was first used therapeutically in the mid 1930s and for flaw detection between 1939 and 1945 (Firestone, 1945; Desch et al., 1946).

Since the Second World War, considerable progress has been made in the development of new piezoelectric crystals, ferroelectric ceramics, and magnetrostrictive materials, and the applications of ultrasound have increased and diversified, particularly in recent years. Fig. 6 includes examples of ultrasound devices used in medicine, industry, consumer products, and signal processing and testing, in relation to ultrasound frequency. Besides the potential for occupational exposure to ultrasound in industrial and medical applications, members of the general population are now exposed to various consumer-oriented devices. However, medical applications are the most rapidly increasing source of exposure. This section includes a brief review of domestic, industrial, commercial, and medical sources and applications of ultrasound.

5.1 Domestic Sources

An ever increasing number of consumer-oriented devices emitting ultrasound are being manufactured. Examples are garage door openers, television channel selectors, remote controls, burglar alarms, dog whistles, bird and rodent scarers, traffic control devices, and range-finders on cameras. In general, low intensities and frequencies at the lower end of the ultrasound range (20-100 kHz) are used in these applications and the ultrasound is usually propagated in air, so that the beam is rapidly attenuated over short distances.

5.2 Industrial and Commercial Sources

The industrial and commercial applications of ultrasound have been reviewed in a number of reports (Lemons & Quate, 1975; Lynnworth, 1975; Shoh, 1975; Jacke, 1979; Repacholi, 1981; Rooney, 1981). Generally, these applications can be

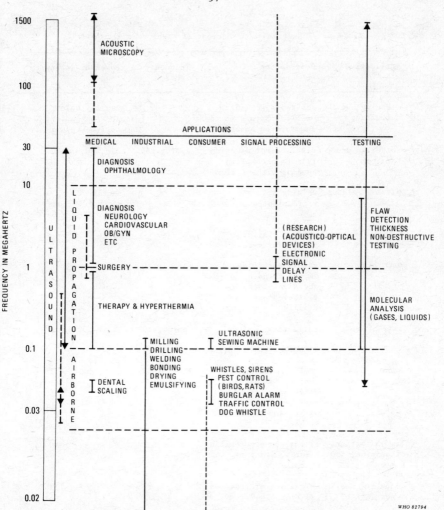

Fig. 6. The ultrasound spectrum. Applications of ultrasound in medicine, industry, consumer products, signal processing and testing, are shown in relation to ultrasound frequency in megahertz (Modified from: IRPA, 1977).

divided into two categories (high- and low-power), depending on the power or intensity levels involved. High-power applications usually rely on compound vibration-induced phenomena occurring in the object or material being irradiated. These phenomena include cavitation and microstreaming in liquids, heating, and droplet formation at liquid/liquid and liquid/gas interfaces. Some of the more common applications of high-power

Table 5. Industrial applications of high-power ultrasound[a]

Application	Description	Frequency (kHz)	Power or intensity range
cleaning and degreasing	cavitating cleaning solution scrubs parts immersed in solution	18 - 100	usually below 10 W/cm^2 but up to 100 W power
soldering and brazing	displacement of oxide film to accomplish bonding without flux	approx. 30	2 - 200 W/cm^2
plastic welding	welding soft and rigid plastic	20 - 60	usually 20 - 30 W/cm^2 but power below 1000 W output
metal welding	welding similar and dissimilar metals	10 - 60	up to 10 000 W/cm^2
machining	rotary machining, impact grinding using abrasive slurry, vibration-assisted drilling	usually 20	
extraction	extracting perfume, juices, chemicals from flowers, fruits, plants	approx. 20	about 500 W/cm^2
atomization	fuel atomization to improve combustion efficiency and reduce pollution; also dispersion of molten metals	20 - 30 000	up to 800 W
emulsification, dispersion, and homogenization	mixing and homogenizing liquids, slurries, and creams	-	-
defoaming and degassing	separation of foam and gas from liquid, reducing gas and foam content	-	-
foaming of beverages	displacing air by foam in bottles or containers prior to capping	-	-
electroplating	increases plating rates and produces denser, more uniform deposit	approx. 27	30 W

Table 5 (contd).

Application	Description	Frequency (kHz)	Power or intensity range
erosion	cavitation erosion testing, deburring, stripping	-	-
drying	drying heat-sensitive powders, foodstuffs, pharmaceuticals	-	-
cutting	cutting small holes in ceramics, glass, and semi-conductors	approx. 20	about 150 W

a From: Repacholi (1981).

ultrasound are described in Table 5 together with the
ultrasound frequency and power or intensity range used, where
these variables are known. The most practical frequency range
for these applications is 20-60 kHz. Most industrial
ultrasound is produced using an electrostrictive or
magnetostrictive transducer (Lynnworth, 1975), in which the
dimensions of the elements change in response to an applied
electric or magnetic field.

Probably the oldest industrial application is cleaning by
means of cavitation and microstreaming mechanisms. Most
cleaning tanks operate at intensities below 10 W/cm^2,
2 W/cm^2 being commonly used.

Plastic welding with ultrasound became popular in the mid
1960s and ultrasound is now used to assemble toys, appliances,
and thermoplastic parts. At frequencies above 20 kHz and
intensities of more than 20 W/cm^2, sufficient heat is
produced to melt the plastic at the required locations. The
principal advantages of this method are speed, cleanliness,
easy automation, and welding in normally inaccessible places.
An interesting application is the ultrasonic sewing machine.
Here woven or nonwoven fibres can be "sewn" together without
thread.

Metal welding was introduced commercially in the late
1950s and is used in the semiconductor industry for welding or
microbonding miniature conductors. The process involves
relatively low temperatures, usually below the melting point

Table 6. Low-power applications of ultrasound in industry[a]

Application	Principle	Frequency
Measurement of:		
flow	determining flow rates for gases, liquids, and solids - Doppler technique	1 - 10 MHz
elastic properties	relating speed of sound to resonance modes of polarization	25 kHz - 300 MHz
temperature	response to temperature dependence of sound, speed, or attenuation	up to 30 MHz
thickness	timing round trip interval of pulse	2 - 10 MHz
density, porosity	resonant and non-resonant probe transmission	up to 50 kHz
grain size of metals	ultrasound attenuation	few MHz
pressure	frequency of quartz crystal resonator changes with applied pressure	0.5 - 1 MHz
level	attenuation of ultrasound beam or measure travel time (pulse echo technique)	around 100 kHz
Counting	beam interruptions counted	40 kHz
Gas leaks	detection of ultrasonic "noise"	36 - 44 kHz
Flaw detection	observe discontinuities in reflected beam	25 kHz to 25 MHz (mW power)
Delay lines	transform electric signal into ultrasound and back again after ultrasound has travelled a well-defined path	few MHz
Burglar alarms	ultrasound beamed into room and a certain level of reflected beam is monitored; if this level changes (with intruder) alarm sounds	18 - 50 kHz (mW powers)
Pest control	frequency and intensity of ultrasound bothersome to pests - inaudible to human beings	18 - 50 kHz (mW powers)
Sonar	Doppler method determines presence and velocity of object	5 - 50 kHz
Acoustic microscope	observe phase shift and attenuation of ultrasound beam by the specimen	100 - 3000 MHz

[a] Adapted from: Lynnworth (1975).

of the metal. The welding depends on ultrasonic cleaning.
Ultrasonic shear causes mutual abrasion of the two surfaces so
that exposed plasticized or metal surfaces can be joined under
pressure to form a "solid-state" bond. For this process, very
high intensities are needed at the welding tip (of the order
of 2000 W/cm^2 at frequencies ranging from 40 to 60 kHz).
Ultrasound soldering, without fluxes, has also been
carried out since the early 1950s. Cavitation in the molten
solder erodes the surface of metal oxides and exposes the
clean metal to the solder. Simultaneous cleaning and tinning
of the metal can be effected using ultrasonic intensities up
to 100 W/cm^2, at frequencies between 20 and 50 kHz.
The machining of metals and ceramics can be carried out
using an abrasive slurry between the vibrating tool and the
work-piece. With a rotary machine and axial ultrasonic
vibration, metals and other hard materials can be machined
using diamond-impregnated core bits. Ultrasonic cavitation
accelerates the cutting action of the water-cooled core bits.
Usually, these devices operate at about 20 kHz.
In high-power applications, the materials being worked are
physically changed, whereas, in low-power applications, the
ultrasound is used to examine rather than alter the materials.
In many cases, low-power applications involve frequencies in
the megahertz range (Table 6). Applications include: the
determination of viscosity, transport properties, position,
phase, composition, anisotropy and texture, grain size, stress
and strain, elastic properties; the detection of bubbles,
particles, and leaks; non-destructive testing; acoustic
emission; imaging and holography; and counting by means of
beam disruptions. Many of the devices used in these
applications have intrusive ultrasonic probes, but
non-invasive pulsed and resonance techniques are also used.

5.2.1 Airborne ultrasound exposure levels

There is not a great deal of information concerning sound
pressure levels produced by devices emitting airborne
ultrasound. The US Bureau of Radiological Health has surveyed
the output of several intrusion devices. Peak sound pressure
levels ranged from 80 dB to 93 dB (centre frequency of
one-third octave band) for those devices emitting at 20 kHz,
85 dB to 100 dB (half octave band levels) for those emitting
at 25 kHz, and 75 dB to 90 dB for those at 16 kHz (Herman &
Powell, 1981). These levels were measured at positions where

people were likely to remain for a reasonable length of time. In some cases, levels were as high as 140 dB at the surface of the radiating transducer.

Michael et al. (1974) monitored the output of several devices, including ultrasonic cleaners. Sound pressure levels measured near some ultrasonic cleaners surveyed were as high as 117 dB (20 kHz centre frequency of one-third octave band). Ultrasonic energy emitted into air from other ultrasonic cleaners of 300 W and 150 W, measured at 1 m from the cleaners, was 127 dB and 113 dB (28 kHz centre frequency one-third octave band), respectively (Ide & Ohira, 1975). Similar results were obtained by Crabtree & Forshaw (1977) and Herman & Powell (1981).

A dental drill emitted approximately 80 dB (one-third octave band sound from 16 kHz to 100 kHz), and an insect repeller radiated 61 dB (16 kHz centre frequency, one-third octave band). More detailed information on emissions of airborne ultrasound from various devices has been compiled by Michael et al. (1974).

5.3 Medical Applications

The use of ultrasound in medicine has grown rapidly since the early 1970s, especially in the diagnostic field. This is the result of the availability of good imaging equipment, the development of many new applications, and the increasingly accurate diagnoses that can be made using new techniques. In addition, there is a common contention that no risks are associated with ultrasound exposure.

In the past, imaging equipment has been generally confined to hospital centres, but today, with the marketing of imaging and Doppler devices at relatively low cost, it is common for obstetricians to have the equipment in their private clinics. In many countries, more than 50% of women are exposed to ultrasound during pregnancy and, in some clinics, all women are examined one or more times.

5.3.1 Diagnosis

Ultrasound was introduced into diagnostic medicine in the mid 1950s and its use has increased at such a rate that "with expanding services in ultrasound diagnosis, the frequency of human exposure is increasing with the potential that essentially the entire population of some countries may be

exposed" (IRPA, 1977). The National Center for Devices and
Radiological Health (US Department of Health and Human
Services) estimates that the availability of equipment will be
such that every pregnant woman in the USA could undergo at
least one ultrasound examination of the fetus (Stewart &
Stratmeyer, 1982).

Most medical diagnostic applications of ultrasound are in
the frequency range of 1-10 MHz, except for ophthalmological
examinations, which may be performed at frequencies up to 30
MHz. These examinations are carried out using either pulsed or
cw irradiation.

Added to the growth in sales of equipment and the
increasing numbers of people being exposed to ultrasound is
the fact that new diagnostic techniques are constantly being
developed. With sophisticated imaging devices, ultrasound
imaging technology is making great advances. Since the
development of computerized axial tomography (Hounsfield,
1973) using X-rays, analogous images have been obtained using
ultrasound. Ultrasonic spectroscopy, time-delay spectrometry,
and holographic techniques all offer new potential for this
expanding imaging modality.

Reviews of the diagnostic applications of ultrasound
include those by Lyons (1982), Repacholi (1981), and
Stephenson & Weaver (1981). Some of the areas of the body
commonly investigated and the types of examination performed
are listed in Table 7. From this compilation of diagnostic
procedures, it can be seen that certain areas of the body are
efficiently examined using ultrasound. Areas better examined
with other imaging modalities are those containing large
amounts of gas (e.g., lungs).

5.3.1.1 Exposure levels from diagnostic ultrasound equipment

While, at present, most manufacturers fail to provide
information on exposure levels with their equipment,
ultrasonic intensity levels and total power output
measurements from commercial diagnostic instruments have been
reported by several investigators (Hill, 1971; Rooney, 1973;
Carson et al., 1978; Farmery & Whittingham, 1978; Kossoff,
1978; Stewart, 1979; Zweifel, 1979). These results should be
interpreted with care, since different criteria and techniques
were employed to obtain the data. Output levels from a limited
number of different types of diagnostic devices, reported by
various investigators, are summarized in Table 8.

The levels of output from cw peripheral vascular Doppler
units are high, compared with those from obstetric Doppler

Table 7.   Some applications of diagnostic ultrasound[a]

Part of interest	Measurement made
1. Head	echoencephalography (head scan and brain scan) for midline position determination and ventricular size
brain	neonatal brain tomographic scans, hydrocephalus evaluation
2. Eyes and orbit	ophthalmic echography (eye scan) for ultrasonic biometry, foreign body localization, mass evaluation, retinal detachment
3. Neck	arterial flow studies, plaque evaluation, carotid artery
thyroid	thyroid echography (thyroid scan) for mass evaluation
4. Chest	
heart	echocardiography (heart scan) for pericardial effusion, valve investigation, wall evaluation (motion, thickness), chamber size and function, tumour detection, intra-cardiac blood flow
pleural space	effusion localization
breast	breast echography (breast scan) for mass evaluation
5. Abdomen	
liver	
kidneys	evaluation of size, parenchyma,
spleen	and associated masses
pancreas	
gallbladder	stone detection
biliary ducts	evaluation of size
aorta	aneurysmal dilatation
peritoneal space	ascites and abscess detection
6. Pelvis	
uterus (pregnant)	evaluation of fetus, gestational sac, estimation of fetal age, diagnosis of multiple pregnancy, placental localization, amniotic cavity, fetal heart monitoring, fetal growth rate, molar pregnancy, ectopic pregnancy, fetal breathing, congenital anomalies
uterus (non-pregnant)	evaluate nature and size of masses
ovaries	following Graafian follicle development for ovulation timing
bladder	tumour assessment
prostate	tumour detection
7. Extremities	
arteries and veins	vascular studies, peripheral flow
8. Ultrasonic guidance procedures	Ultrasonic guidance for amniocentesis, needle biopsy, thoracentesis or cyst location, placement of ionizing radiation therapy fields

[a] From: Lyons (1982).

Table 8. Range of output intensities found in beams produced by medical ultrasonic equipment[a]

Type of equipment	Spatial average, temporal average (SATA) intensity on the radiating surface	Spatial peak temporal average (SPTA) intensity	Spatial peak pulse average (SPPA) intensity	Spatial peak temporal peak (SPTP) intensity
static pulse echo scanners A-mode	0.2–20 mW/cm²	0.6–125 mW/cm²	0.1–160 W/cm²	0.4–1000 W/cm²
automatic sector scanners (phased arrays and wobblers)	0.5–60 mW/cm²	2–200 mW/cm²	0.3–100 W/cm²	4–120 W/cm²
sequenced linear arrays	0.06–10 mW/cm²	0.1–12 mW/cm²	0.3–100 W/cm²	4–120 W/cm²
pulsed Doppler, primarily for cardiac work	3–32 mW/cm²	20–290 mW/cm²	1–14 W/cm²	2–28 W/cm²
Doppler instruments, primarily for obstetric applications	0.26–25 mW/cm²	0.75–75 mW/cm²		
continuous wave Doppler, primarily for peripheral vascular investigations	10–400 mW/cm²	20–800 mW/cm²		
therapy continuous wave	up to 4 W/cm²	0–16 W/cm²		
therapy, gated mode	up to 1 W/cm²	0–4 W/cm²		

[a] Intensity data were obtained from published values in the literature (Rooney, 1973; Etienne et al., 1976; Carson et al., 1978; O'Brien, 1978; Nyborg, 1979; Stewart, 1979; AIUM-NEMA, 1981; Hill & ter Haar, 1981; Stewart & Stratmeyer, 1982). Measurements were made with transducers immersed in water.

units. This is due, in part, to the sensitivity that is required to detect the small signals received from flowing blood. The SATA intensity output levels at the face of the transducer for single element pulse echo A and B mode imaging units are in the low mW/cm^2 range. The intensities at the transducer face are much lower than the intensities measured at the focal distance for units using focusing transducers. Though the reported SATA intensities may be in the mW/cm^2 range (Table 8), the SPTP intensities can sometimes be in the hundreds of W/cm^2 range.

In the case of automatic scanners equipped with a mechanical sector scan or a multi-element transducer providing a linear or sector scan motion of the ultrasound beam, the time pattern of the sound field at a point of interest is characterized by the pulse shape and pulse duration (typically around 1 µs), the pulse repetition frequency (typically a few kHz) and the frame repetition frequency (typically 10-50 Hz). When the beam is scanned over the point of interest, a short sequence of pulses, the number of which is given by the ratio of the beam width to the beam shift between subsequent pulses (typically 2-5 pulses) is recorded at this point. While SPTP intensities of the order of 10 W/cm^2 occur at the pressure maxima of these few pulses, the SPTA intensity, when averaged over the short sequence of pulses, is of the order of 1-10 mW/cm^2. After the short pulse group, the ultrasound intensity at the point of interest remains at a very low level while the beam is scanned to other positions. Thus the SPTA intensity, when averaged over the total period of one frame, is proportional to the ratio of the number of pulses in the short sequence to the total number of pulses per frame. This ratio may vary from 0.01 to 0.05, so that SPTA intensities of 0.01-0.5 mW/cm^2 result, when averaged over the total frame time.

5.3.2 Therapy

Ultrasound therapy usually involves the application of a hand-held ultrasound transducer to the injured area of a patient, and treatment with either a cw or pulsed beam. Intensities employed in physiotherapy normally range from about 100 mW/cm^2 to 3 W/cm^2. The transducer head is generally moved over the area of injury to obtain as uniform a treatment distribution as possible.

Lehmann et al. (1974, 1978) pointed out that the main therapeutic value of ultrasound was related to its selectivity of absorption. In soft tissue, this absorption may be directly related to the protein content of the tissue (Piersol et al., 1952; Bamber et al., 1981). Lehmann et al. (1974) also claimed that the benefit of ultrasound as a therapeutic agent was that it heated selectively the areas that required heating, including superficial bone, scar tissue within soft tissue, tendons and tendon sheaths, etc. Furthermore, they claimed that ultrasound might accelerate the diffusion process across biological membranes, implying an increased rate of healing. There may also be low-intensity, ultrasound-induced, non-thermal effects, which may be important in certain physiotherapeutic applications, such as the breakdown of fibrous adhesions at the site of a surgical incision (Wells, 1977; Coakley, 1978; ter Haar et al., 1980).

The stimulatory effect of ultrasound in healing ulcers in human subjects has been reported by various investigators (Dyson et al., 1976; Goralčuk & Košik, 1976). Dyson et al., (1976) suggested that nonthermal mechanisms might be involved in the beneficial therapeutic action of ultrasound on tissues.

It is, however, very difficult to assess the benefits from ultrasound therapy, as Roman (1960) found. Of 100 patients treated or sham-irradiated for lower back pains, bursitis of the shoulder, and myalgia, 60% receiving ultrasound were categorized as normal, but 72% of the shams were in the same category. Many more well-controlled studies ought to be conducted to identify optimal exposure conditions and to eliminate ineffective treatments.

5.3.2.1 Exposure levels from therapeutic ultrasound equipment

Ultrasonic therapy units are usually equipped with an indicator of the total output power (either a meter or calibrated dial), a timer, and a power output adjustment. They usually register total output power in watts (W) and intensity in W/cm^2, which is the power divided by the effective radiating area of the transducer. Some ultrasound units can be operated in either cw or gated mode (Fig. 2). In the gated mode, most units operate at a gate repetition rate from about 8 Hz to 120 Hz with a gate width of up to 12 ms. Gated mode therapy units are normally calibrated in terms of the cycle average intensity (I_a) (Appendix I).

In cw operation, the ultrasonic power and spatial average intensity can be adjusted up to about 20 watts and 4.0 W/cm^2, respectively (Repacholi & Benwell, 1979). In gated mode, the peak power and temporal peak spatial average intensity in one unit could be adjusted up to approximately 80 watts and 8.0 W/cm^2, respectively (Stewart et al., 1982).

Because beam divergence is a function of applicator size for a given ultrasonic frequency, therapy transducers with beam areas of less than 5 cm^2 have been stated by some to be unacceptable (Lehmann, 1965a, b). In addition, with a small beam it may be difficult to treat a large area on an individual. On the other hand, if the radiating area of the applicator is too large, it may be difficult to maintain contact with curved surfaces of the body during treatment. The effective radiating area of therapy applicators generally ranges between 1 and 10 cm^2.

5.3.3 Surgical applications

Ultrasound has been used in vestibular surgery for the treatment of Ménière's disease. The treatment involves ultrasound exposure of the vestibular end organ to SPTA intensities of 10-22 W/cm^2 from a specially designed ultrasonic probe (James, 1963; Kossoff & Khan, 1966; Sorensen & Andersen, 1976).

Kelman (1967) first described the use of a phaco-emulsification and aspiration technique for the removal of cataracts in situ. The low-frequency probe (phacoemulsifier) is inserted into the lens of the eye to break up the cataract, then the broken pieces are sucked out through a hollow tube. This technique has been refined and used successfully (Emery, 1974; Emery et al., 1974; Emery & Paton, 1974; Girard, 1974).

Other surgical procedures in which ultrasound has been used include: cleaning of obstructed blood vessels and ureters, and fragmenting kidney-stones (Davies et al., 1974, 1977; Stumpff et al., 1975; Finkler & Hausler, 1976; Yeas & Barnes, 1970), neurosurgery (Arslan et al., 1973), and cutting and welding tissues (Goliamina, 1974; Hodgson et al., 1979; Williams & Hodgson, 1979).

Non-surgical destruction of kidney-stones can be performed by repeated application of acoustic shock-waves (Chaussy et al., 1980). The patient is treated lying in a water-bath, where high-intensity ultrasound pulses of microsecond duration, are generated by electrical discharges from a spark-gap, placed in one focus of a concentrating ellipsoidal ultrasound mirror system. Exact positioning of the patient is performed under X-ray guidance. This enables continuous visualization of the gradual disintegration of the stone during the treatment.

5.3.4 Other medical applications

Ultrasound has been used to atomize liquids, in order to produce aerosols that can maintain a humid atmosphere in a ventilating assistor (Miller et al., 1968). Boucher & Krueter (1968) described several ultrasonic nebulizers which are available commercially. These devices operate at 1-1.4 MHz and produce aerosols with particle diameters of between 1 and 1.4 μm.

Methods in which gas bubbles are detected by increases in ultrasound attenuation due to the bubbles in tissue have been described by Manley (1969). In other methods, the fact that gas bubbles circulating in vivo give rise to characteristic changes in the output from a cw Doppler device has been used to detect these bubbles (Evans & Walder, 1970). Ultrasound frequencies ranging from 1 to 3 MHz and intensities of a few mW/cm^2 are employed in these procedures. Ultrasonic pulse-echo imaging has also been used to study decompression-induced gas bubbles in vivo (Daniels et al., 1979).

The application of ultrasound to the acupuncture meridian system has been reported by Khoe (1977). Output powers of 0.25-1 W for 0.5-2 min are used at each acupuncture point. Presumably, the frequency of the transducer is somewhere in the range of 0.8-3 MHz, though this is not specifically mentioned by the author. This technique was claimed to be effective for a variety of viral, bacterial, and fungal diseases; allergic, gastrointestinal, gynaecological, and musculo-skeletal disorders; and cardiovascular diseases.

Kremkau (1979) has completed a review of events leading up to the relatively new use of ultrasound for cancer therapy. Ultrasound can produce hyperthermia in surface and deep-seated tissue volumes (Lele, 1967; Palzer & Heidelburger, 1973) (section 6.4.6.5).

5.3.5 Dentistry

The ultrasonic drill was developed in the early 1960s but never really gained acceptance in dentistry because of the introduction of the high-speed rotary drill. However, the number of other applications of ultrasound in dentistry has been steadily growing (Balamuth, 1967). These include cleaning and calculus removal, gingivectomy, root canal reaming, ortho-dontic filling, amalgam packing, and gold-foil manipulation. Conventional techniques for these tasks are fairly satis-factory, but there is no doubt that the silence and ease of the ultrasonic methods relieves the patient of some of the stress associated with dental treatment. Frost (1977) estimated that in the USA there may be as many as 100 000 ultrasonic units in use in dental offices for scaling teeth and peridontal care.

It appears that long-term studies on the biological effects of ultrasound devices in dentistry have not been reported in the literature. The extent to which these devices are hazardous depends largely on how they are used. While investigators tend to attribute most of the bioeffects to heating, the cavitation associated with the water coolant spray cannot be ignored, especially subgingivally. When used improperly, ultrasound dental devices are apparently more likely to be hazardous or ineffective than conventional techniques. Most of the commonly used dental devices operate in the frequency range of 20-40 kHz.

6. EFFECTS OF ULTRASOUND ON BIOLOGICAL SYSTEMS

6.1 Introduction

The studies reviewed in this section have been arranged according to the complexity of the biological systems under study, i.e., from macromolecules to complete multicellular organisms. Caution must be exercised, when interpreting the results of many of the studies involving macromolecules and cells in suspension. The acoustic mechanism(s) of interaction predominantly responsible for effects in these systems may not necessarily be the same as those responsible for effects in intact tissue or organisms. However, because of the problems inherent in using intact animals to search for unpredicted effects, macromolecular and cellular studies may provide valuable information concerning end-points that might reasonably be examined in higher level organisms.

The data concerning biological effects are incomplete, because few biological structures have been subjected to systematic examination for effects from ultrasound. Estimates of ultrasound field variables in living systems still suffer from a lack of accepted methods of measurement, and often from inadequately stated experimental conditions. In many in vitro experiments, cell suspensions have been in contact with foreign surfaces (e.g., test-tubes, culture dishes, plastic) during ultrasound exposure. The complex acoustic fields reflected from these surfaces frequently make it difficult to determine the cell exposure levels and to compare the results with those of studies conducted using different experimental arrangements.

Unfortunately, the SATA intensity has been determined in different ways in many bioeffects reports. In some studies, it has been determined as indicated in Appendix II. In others, the total power of the beam has been determined and divided by the area of the transducer face. This variation in the methods of determination of SATA intensity introduces difficulties when comparing the results of different laboratories.

The evidence that is presented should be considered as inconclusive, in most cases, until confirmed by independent laboratories.

6.2 Biological Molecules

Extensive work has been carried out on the action of ultrasound on chemical systems and, in particular, on large molecules of biological interest (El'piner, 1964). The effects at this level are broadly of three kinds (Edmonds, 1972): (a) passive absorption of the (coherent) ultrasound energy; (b) mechanical degradation of large molecules; and (c) chemical effects, apparently attributable to the action of cavitation in releasing chemically active "free radical" species in irradiated solutions.

It has been shown that the absorption properties of blood are mainly determined by, and are directly proportional to, its protein content (Kremkau & Carstensen, 1972; O'Brien & Dunn, 1972). Furthermore, since the frequency dependence of ultrasound absorption by whole and homogenized liver tissue is very similar, it has been concluded that approximately two-thirds of the absorption occurs at the macromolecular level, with one-third due to the tissue structure (O'Brien & Dunn, 1972). For a more extensive coverage of the literature in this area, the reader is referred to reviews by Repacholi (1981) and Stewart & Stratmeyer (1982).

There have been a number of studies on the effects of ultrasound on solutions of purified DNA. Hill et al. (1969) found that a 3-min exposure of calf thymus DNA to cw 1 MHz ultrasound at 400 mW/cm^2 resulted in DNA degradation. Similarly, Galperin-Lemaitre et al. (1975) reported that exposing calf thymus DNA to 1 MHz ultrasound, at 200 mW/cm^2, resulted in DNA degradation. The DNA strand breakage was thought to be due to hydrodynamic shear stress generated by acoustic cavitational activity.

In summary, though solutions of macromolecules such as proteins and nucleic acids are capable of absorbing ultrasound in the megahertz frequency range, damage has usually been reported only as a result of cavitation. However, it is not clear if these data can be extrapolated to the in vivo situation, since the structure of DNA in solution bears little resemblance to its structure in vivo.

6.3 Cells

Studies aimed at elucidating the mechanisms of action of a particular agent may be more readily performed and analysed using cell suspensions than the whole animal, because of the absence of numerous uncontrollable biological variables. Effects observed in mammalian cells, after ultrasound exposure, include: modification of macromolecular synthetic pathways and cellular ultrastructure; cell lysis, cellular inactivation, and altered growth properties; and chromosomal changes. Current information concerning such effects will be discussed in this section with the exception of chromosomal changes, which will be discussed in section 6.4.4.

6.3.1 Effects on macromolecular synthesis and ultrastructure

Alterations in the rates of protein and DNA synthesis have been reported to occur in cells grown in tissue culture, when exposed to ultrasound.

6.3.1.1 Protein synthesis

Stimulation of the rate of protein synthesis was observed 4 days after exposure of human fibroblasts for 5 min to cw 3 MHz ultrasound at intensities of 0.5-2.0 W/cm² (Harvey et al., 1975). Continuous wave exposure at 0.5 W/cm² caused total protein synthesis in fibroblasts to increase by 20%, while exposure to pulsed ultrasound (pulse duration 2 ms; duty factor, 0.2) at the same average intensity resulted in a 30% increase compared with control values (Harvey et al., 1975; Webster et al., 1978). The stimulation, which appeared to be inversely related to the ultrasound frequency in the range 1-5 MHz, did not occur when the cells were pretreated with cortisol. The authors suggested that the increased protein synthesis observed was due to damage to the lysosomal and plasma membranes (possibly by a cavitational mechanism of action), since no ultrastructural changes occurred if the cells were exposed at elevated pressures.

Belewa-Staikowa & Kraschkowa (1967) observed an increase in protein synthesis in hepatic, renal, and myocardial tissue treated with a single, 5-min exposure to a therapy transducer at intensities of both 0.2 and 0.6 W/cm². However, protein synthesis was retarded at 1 W/cm². A similar effect was found by Repacholi (1982) in that stimulation of protein synthesis occurred in human lymphocytes at low cw therapeutic intensities (870 kHz, 1.1 W/cm², 30 min), and retardation at higher intensities (3-4 W/cm²).

6.3.1.2 DNA

Increased DNA synthesis in vitro was observed 1, 2, and 3 days after exposure of excised neonatal mouse tibiae to cw 1 MHz ultrasound at 1.8 W/cm² (Elmer & Fleischer, 1974).

However, no statistically significant differences were
observed in either protein accumulation or in bone elongation
compared with the controls.

Levels of (^{3}H) thymidine and (^{3}H) deoxyuridine
incorporated into DNA decreased to 54% and 42% of control
values, respectively, following exposure of mouse leukaemia
1210 cells to 2.22 MHz ultrasound for 10 min, at a mean
spatial intensity of 10 W/cm^2 (Kaufman & Kremkau, 1978). The
authors found that ultrasound caused reversible injury in the
cell, which was not readily reversed in the presence of
cytotoxic drugs, and that this resulted in a significant
decrease in the lethal potential of the leukaemia cells. A
significant immediate inhibition in the incorporation of
(^{3}H) thymidine was also found by Repacholi et al. (1979) and
Repacholi (1982), when human blood lymphocytes were exposed in
vitro to therapeutic ultrasound (cw near-field, 870 kHz, 4
W/cm^2, for 30 min). The uptake of the radioactive precursors
returned to control levels, 2-3 days after exposure
(Repacholi, 1981).

Fung et al. (1978) exposed activated human lymphocytes to
cw ultrasound for 0-30 min using a commercial fetal Doppler
unit. The uptake of (^{3}H) thymidine over an 18-h period, 1
day after ultrasound exposure, was found to be biphasic. There
were lymphocytes that showed significant stimulation in uptake
at short exposure times (3-12-min exposure) with a return to
control values at longer exposure times (15-30-min exposure),
and lymphocytes that did not exhibit any stimulatory effect at
short exposure times, but showed a significant reduction in
uptake with 12- and 30-min exposures.

In a study by Liebeskind et al. (1979a), exposure of
synchronized HeLa cells in culture to pulsed 2.5 MHz
ultrasound at a SATA intensity of 17 mW/cm^2 (35.4 W/cm^2
SPTP intensity) induced unscheduled, non-S-phase (repair) DNA
synthesis. This result suggested that the DNA had been damaged
by the ultrasonic exposure. A similar effect was reported by
Repacholi & Kaplan (1980), who found non-S-phase unscheduled
DNA synthesis in human peripheral blood lymphocytes exposed to
cw near-field, 870 kHz ultrasound at 4 W/cm^2 for 30 min.

In another study, Liebeskind et al. (1979b) found a small
but significant increase in the frequency of sister chromatid
exchanges (SCE), following a 30-min exposure of normal human
lymphocytes to pulsed diagnostic ultrasound of frequency
2.0 MHz, at 2.7 and 5.0 mW/cm^2 (SATA intensity). Results
consistent with these were reported by Haupt et al. (1981) who
used a commercial real time scanner, having a pulse repetition
frequency of 2420 Hz at 3.5 MHz, pulse duration of 0.89 μs,

estimated SPTP intensity of 2 W/cm², and SPTA intensity of
0.02 mW/cm² for 7.5-90 min. However, Morris et al. (1978),
who used cw 1 MHz ultrasound exposures at intensities of 9.1,
15.3, 27, and 36 W/cm² did not find an increase in SCEs. The
time of exposure was also different in that unstimulated
stationary phase (Go) lymphocytes were exposed before both
divisions, whereas, in the studies by Liebeskind et al. and
Haupt et al., stimulated lymphocytes were exposed after the
first division, but before the second. Thus the experimental
conditions were completely different; the cells used by Morris
et al. (1978) were in a less sensitive state and therefore the
results are not comparable. Wegner et al. (1980), who exposed
Chinese hamster ovary cells to cw 2.2 MHz ultrasound at 10
mW/cm² for 30 and 90 min using a fetal Doppler unit, also
did not observe any increase in SCE. These data raise
questions about the possible effectiveness of pulsed diagnostic ultrasound compared with cw exposures in causing SCE.

The significance of SCE in relation to biological hazard
is not understood, though the phenomenon is generally held to
be undesirable. For some other types of insults, sister
chromatid assay has been suggested to be a sensitive measure
of genetic damage, because the frequency of exchanges
increases after exposure of cells to known mutagens and
carcinogens (Stetka & Wolff, 1977). The SCE method has been
advocated as a direct test of mutagenic or carcinogenic agents
(Latt & Schreck, 1980; Shiraishi & Sandberg, 1980).

6.3.1.3 Cell membrane

Ultrasonically-induced functional alterations in the
plasma membrane have been reported by a number of investigators. These alterations include increased permeability,
decreased active transport, decreased non-mediated transport,
and decreased electrophoretic mobility. A 5% decrease in the
non-mediated transport of leucine in avian erythrocytes
following a 30-min, 1 MHz ultrasound exposure at an intensity
of 0.6 W/cm² was reported by Bundy et al. (1978). However,
no change was observed in the active transport of (^{3}H)
thymidine in human lymphocytes exposed to cw 870 kHz
ultrasound at intensities up to 4 W/cm², for 30 min
(Repacholi, 1982).

A reduction in the electrophoretic mobility of Ehrlich
ascites tumour cells observed by Repacholi (1970) and
Repacholi et al. (1971) was directly proportional to the
square root of the ultrasonic frequency used in the range of
0.5-3.2 MHz (Taylor & Newman, 1972). This reduction in
mobility was reported to be independent of the pulse length

over the range of 20 μs–10 ms (peak intensity was 10 W/cm^2; duty factor, 0.1, exposure time, 5 min). The change in mobility was presumably a result of alteration of the surface charge of the cells. This effect was also reported by Joshi et al. (1973) and later reported to be reversible and non-lethal by Hill & ter Haar (1981).

A mechanical stress mechanism of action was suggested to be the cause of an increase in the permeability of human erythrocyte membranes to potassium ions, observed following ultrasound exposure in vitro for 5–30 min (1 MHz, 0.5–3.0 W/cm^2) (Lota & Darling, 1955). A decrease in potassium content was reported to occur following sonication of rat thymocytes for 40 min, using an ultrasonic therapy unit operated at 3 MHz and 2 W/cm^2 (Chapman et al., 1980). These changes appeared to be a result of both a decreased influx and an increased efflux of potassium.

Changes in the concentrations of membrane-associated cAMP and cGMP have profound effects on a wide variety of cellular processes. However, no alterations in the amount of cAMP and cGMP could be detected following exposure of human amniotic cells or mouse peritoneal cells to cw 1 MHz ultrasound at 1 W/cm^2 for 33 min (Glick et al., 1979).

Siegel et al. (1979) reported that dispersed cultured human cells seeded in plastic Petri dishes showed significantly reduced cellular attachment after 0.5 min of exposure to a pulsed, 2.25 MHz clinical diagnostic ultrasound source (approximate SATA intensity, 10 mW/cm^2). The authors suggested that, if cellular attachment were to be altered in vivo, it could affect implantation, morphogenesis, and development. These results may be related to findings described by Liebeskind et al. (1981a) on the spectacular morphological changes in cell surface characteristics observed after pulsed diagnostic ultrasound exposure. Mouse 3T3 cells examined for up to 37 days after a single exposure demonstrated abnormally large numbers of microvilli and cell projections. Thirty-seven days represents 50 generations for this cell line and suggests that the altered cell surface characteristics were a result of a hereditary change. However, Mummery (1978) did not observe these changes following exposure of fibroblasts to either pulsed or cw therapeutic ultrasound.

Martins (1971) reported that scanning electron micrographs of M3-1 cells exposed to 1 MHz ultrasound at 1.0 and 0.25 W/cm^2 showed a characteristic bumpy outer surface, compared with the smooth outer surface of unexposed cells.

The motility in vitro of sparse populations of human embryo lung fibroblasts was found to increase after exposure to 3 MHz ultrasound at SPTP intensities of 0.5–2.0 W/cm^2, pulsed 2 ms on, 8 ms off for 20 min. This was the result of an increase in directionality rather than an increase in mean speed (Mummery, 1978). The author suggested that this effect could be implicated in the beneficial therapeutic actions of ultrasound on wound healing.

An increase in the calcium ion content of human embryonic lung fibroblasts resulted from in vitro exposure to 3 MHz ultrasound, at SPTP intensities of 2 and 4 W/cm^2 pulsed 2 ms on, 8 ms off, for 20 min. The effect was still observed, when the cells were washed with ethylene diamine tetracetic acid (EDTA) after treatment, but was suppressed by doubling the ambient pressure during sonication. This strongly implicates acoustic cavitation as the dominant mechanism (Mummery, 1978).

In summary, there are several reports indicating that diagnostic levels of pulsed ultrasound can cause structural and functional changes in cell surface characteristics. Because of the importance of the cell surface in immune determination, receptor topography carrier systems, and cell-cell recognition, these changes could have quite important ramifications in vivo. However, the interpretation of the results of cell culture experiments in terms of an in vivo situation is speculative, because of the difficulty in bridging the gap between experimental in vitro work and biological effects that occur in the patient.

6.3.1.4 Intracellular ultrastructural changes

Numerous reports have appeared describing ultrastructural damage to cells exposed to ultrasound. Rat bone-marrow cells in suspension, irradiated with 0.8 MHz ultrasound for 1 min at 1.5 W/cm^2, exhibited gross damage, when examined by electronmicroscopy (Dunn & Coakley, 1972).

Electron microscopic examination of human fibroblasts, irradiated with pulsed, 3 MHz ultrasound at an SATP intensity of 0.5 W/cm^2 (duty factor 0.2), revealed more free ribosomes, increased dilation of the rough endoplasmic reticulum, increased damage to mitochondria and to lysosmal membranes, and more cytoplasmic vacuolation (Harvey et al., 1975). Exposure of HeLa cells to 0.75 MHz ultrasound at an intensity of 0.9 W/cm^2 for 20-120 s caused slits in the cells, holes in the nuclear membranes, separation of the inner and outer nuclear membranes, increase in cell debris, exploded mitochondria, and lesions of the endoplasmic reticulum (Watmough et al., 1977). The results suggested that some of the damage, such as rupture of the nuclear and plasma membranes, may have been due to shear stresses resulting from microstreaming around oscillating microbubbles.

Table 9. Ultrastructural changes following in vivo
exposure to ultrasound

SATA intensity (mW/cm^2)		Total exposure time (min)	Effect observed	Reference
100	(cw)	15	damage to luminal aspect of plasma membrane, cell debris (chick embryo)	Dyson et al. (1974)
1000	--	10	membrane changes, swollen mitochondria, cell debris (rat testes)	Dumontier et al. (1977)
1000	(cw)	9.1	changes in mitochondria (mouse liver, pancreas, kidney)	Stephens et al. (1978)
1000	(cw)	10	membrane changes, changes to mitochondria (germinating spores of Rhizopus nigricans)	Hrazdira & Havelkova (1966)
1000	(cw)	20	swollen basal labyrinth, microvilli, & mitochondria (dog kidney)	Pinčuk et al. (1971)
2000	(cw)	1	necrosis, haemorrhage (mouse liver)	Valtonen (1967)
2500	(cw)	5	vacuolation, necrosis, desquamation, and mural thrombosis (rabbit arteries)	Fallon et al. (1973)
3000	(cw)	5 (multiple)	increase in lysosome destruction (rat liver)	Majewski et al. (1966)
3000	(cw)	5	increase in lysosome destruction (rabbit liver)	Jankowiak & Majewski (1966)
3500	(cw)	3	necrosis, intracytoplasmic vacuolation, destroyed mitochondria (rabbit larynx)	Karduck & Wehmer (1974)

Cachon et al. (1981) conducted studies on the microtubule system of a Heliozoan, using a commercial pulsed diagnostic device emitting 2.5 mW/cm² for 10-20 s at 5 MHz. The microtubules became disorganized within their axopods after exposure to ultrasound and the organisms stopped moving and died rapidly. Electronmicroscopic examination of human blood lymphocytes exposed for 30 min to cw 870 kHz ultrasound at 4 W/cm² also revealed disruption of microtubule formation (Repacholi, 1982).

Results of studies on human lymphocytes and Erlich ascites carcinoma cells suggested a possible disturbance of the mitotic spindle at metaphase following ultrasound exposure (Schnitzler, 1972). Clarke & Hill (1970) reported that, in L51784 cells, the susceptibility to ultrasonic disintegration

Table 10. Ultrastructural changes following in vitro
exposure to ultrasound

SATA intensity (mW/cm²)		Total exposure time (min)	Effect observed	Reference
15	(p)	30	ultrastructural changes (3T3 fibroblast cells & rat peritoneal fluid cells)	Liebeskind et al. (1981b)
15	(p)	30	increase in number of microvilli (mouse 3T3 cells)	Liebeskind et al. (1981a)
500	(p)	5	damage to lysosomes, mitochondria, cytoplasmic vacuoles (human fibroblasts)	Harvey et al. (1975)
800	(cw)	5	increased platelet aggregation (human blood)	Chater & Williams (1977)
900	(cw)	0.3-2	damaged plasma & nuclear membranes, increased cell debris (HeLa cells)	Watmough et al. (1977)
2000	(cw)	2	rupture of myofibrils (chicken muscle)	Samosudova & El'piner (1966)
2600	(cw)	40	deformed erythrocytes (human blood)	Koh (1981)

increased during mitosis. It was suggested that cells are particularly susceptible to damage by ultrasound during mitosis, because major changes in the cell membrane and in internal structure occur during this phase of the cell cycle.

When a 3T3 fibroblast cell line and normal rat peritoneal fluid cells were exposed to pulsed 2 MHz ultrasound at 15 mW/cm^2 for 30 min post-sonication ultrastructural changes were observed (Liebeskind et al., 1981b). The authors concluded that low-intensity, pulsed ultrasound could alter both cellular ultrastructure and metabolism. They suggested that the persistence of disturbances in cell motility, many generations after sonication in vitro, is especially important and it can be speculated that, if fetal cells were to be subtly damaged, it might affect cell migration during organogenesis.

Results of in vivo studies designed to observe cell membrane and intracellular changes (Tables 9 and 10) have, in general, been the same as those of in vitro studies. Mitochondria appear to be some of the intracellular organelles most sensitive to ultrasound exposure, exhibiting swelling, loss of cristae, and eventual disruption of the outer membrane. The endoplasmic reticulum seems to be less sensitive to ultrasound exposure than mitochondria, but, with increasing exposure times, dilation of the cisternae, loss of surface ribosomes, and vesiculation occurs. Most cell damage from sublethal exposures appears to be reparable within four days; however, changes in the mitochondria persist for longer periods of time and may be irreversible (Stephens et al., 1978).

6.3.1.5 Summary

In summary, exposure to ultrasound can cause changes in the ultrastructure of cells in culture, which lead to disruptions in macromolecular synthetic pathways. Certain structural components may be susceptible to damage; these include the nuclear, lysosomal, and plasma membranes, microtubules, the mitotic spindle, and the endoplasmic reticulum. Both ultrastructural and functional changes in the plasma membrane have been reported following exposure to relatively low-intensity pulsed ultrasound. Because of the importance of the cell surface in such functions as immune determination, receptor topography carrier systems, and cell-cell recognition, these changes could have quite important ramifications in vivo.

Though cavitation appears to be the dominant mechanism responsible for many of the ultrasonically-induced structural changes, it seems possible that some of these effects could be caused by noncavitational mechanical stresses. The high acoustic intensities associated with pulsed ultrasound may be of importance in the effects observed. The interpretation of the reported effects of pulsed ultrasound exposure on SCE production in vitro and its possible application to in vivo situations is not known.

6.3.2 Effects of ultrasound on mammalian cell survival and proliferation

Ultrasound at sufficiently high intensities can generate cavitational activity that completely destroys microorganisms, viruses, bacteria, and animal and plant cells (Kato, 1969; Clarke & Hill, 1970; Coakley et al., 1971; Hill, 1972a, b; Kishi et al., 1975; Kaufman et al., 1977; Li et al., 1977; Moore & Coakley, 1977). Ultrasonic disruption of cells at high intensities has also been demonstrated, both in vitro and in vivo (Fry et al., 1970; Taylor & Pond, 1970, 1972; Dunn & Fry, 1971; Lele & Pierce, 1972).

Many studies concerning the cellular effects of ultrasound have had qualitative biological end-points such as cell lysis or morphological changes in cell structure. From the mid-1970s, however, investigators began to focus their attention on quantifiable biological variables such as cell survival and proliferative capacity. Lysis of mouse lymphoma cells in suspension, at ultrasound frequencies and intensities used in clinical medicine, has been documented and correlated with acoustic cavitation (Coakley et al., 1971). Maeda & Murao (1977) found significant growth suppression in human amniotic cells in culture exposed to cw 2 MHz ultrasound at intensities higher than 0.8 W/cm^2 for 1 h. Maeda & Tsuzaki (1981) also observed growth suppression in cultured human amniotic cells exposed to pulsed, 2 MHz ultrasound at SATA intensities higher than 60 mW/cm^2 (1 kHz pulse repetition rate, 3-µs duration, 80 W/cm^2 SPTP intensity).

The importance of peak pulse intensities and other parameters, such as pulse duration and pulse repetition frequency, has been reported by other investigators (Barnett, 1979; Sarvazyan et al., 1980). It has been suggested that intact cells surviving ultrasound exposure remain unaffected, in terms of subsequent growth and proliferation rates (Clarke & Hill, 1969). However, other studies have shown that many of the intact nonlysed cells remaining after ultrasound exposure

6

of mammalian cells in suspension are non-viable, as determined by both vital dye exclusion and colony-forming ability (Kaufman et al., 1977).

Exposure of HeLa and CHO cells for 2-5 min to cw 1 MHz ultrasound resulted in a threshold for cell lysis at an intensity of approximately 1 W/cm^2, with the maximum effects occurring at an intensity of 10 W/cm^2 (Kaufman et al., 1977). Colonies formed from sonicated cells contained fewer cells and a higher frequency of giant cells than colonies formed from appropriate controls (Miller et al., 1977).

Kremkau & Witcofski (1974) reported a significant reduction in the rate of occurrence of mitotic cells in surgically stimulated rat liver exposed in vivo to cw 1.9 MHz ultrasound at an intensity of 60 mW/cm^2. However, Miller et al., (1976a) were unable to confirm these findings with the same biological system exposed for 1 and 5 min to 2.2 MHz ultrasound at intensities in the range of 0.06-16 W/cm^2. One possible explanation for the differences in the results obtained in these studies was that the second method involved a circular motion of the transducer over the animal's ventral surface, while the transducer was kept stationary in the first case. Negative results were also obtained by Barnett & Kossoff (1977), when they exposed regenerating rat liver to pulsed, 2.5 MHz ultrasound, 10-50 kHz pulse repetition rate and a temporal peak intensity of 33 W/cm^2.

Ultrasound exposure of cells in suspension has been shown to induce both immediate and delayed effects (Kaufman & Miller, 1978). Studies performed at elevated temperatures showed that immediate cell lysis was independent of temperature (up to 43 °C), whereas cellular inactivation (as measured by a reduction in plating efficiency) was temperature dependent (Li et al., 1977). These studies indicate that immediate cell death may be caused by large-scale cellular damage (probably resulting from some form of cavitational activity), whereas the delayed effects depend on the cell's ability to repair sublethal damage. These repair mechanisms are less efficient at elevated temperatures.

It appears that there is quite a wide range of "threshold intensities" for the lysis of isolated cells in suspension. Variables contributing to this wide variation include: the gas content of the medium; exposure geometry; ultrasound exposure parameters; and the number and availability of cavitation nuclei. In any given medium, the last of these factors depends critically on the treatment of the medium immediately prior to exposure and the degree of agitation during exposure (Williams, 1982a).

6.3.3 Synergistic effects

Variable results have been obtained following combined exposure to ultrasound and X-rays, including: increases in cell death; increases in chromosomal aberration; reduction in the ionizing radiation dose needed to achieve tumour remission; and increases in cell membrane effects. As an example of divergent results, Todd & Schroy (1974) reported that ultrasound (920 kHz, 0.14 W/cm^2), administered within 10 min of X-irradiation, decreased the dose of 50 kVp X-rays required to prevent 99% of cultured Chinese hamster cells from forming colonies. In contrast, exposure of L5178Y mouse lymphoma cells in suspension to ultrasound did not have any significant effect on the survival of these tumour cells, either alone or by altering the response to X-rays (Clarke et al., 1970). Kunze-Muhl (1981) treated human lymphocytes with cw ultrasound at 20 mW/cm^2 and 3 W/cm^2 and also 20 mW/cm^2 in combination with X-ray exposure, and observed variable increases in chromosomal aberration frequency depending on whether the ultrasound was given before or after X-irradiation.

In a preliminary communication, Burr et al. (1978) reported a highly significant (P<0.00001) relative increase in the number of chromosome aberrations observed in human lymphocytes in vitro when ultrasound was administered at the same time as, or immediately after, 2 Gy of Gamma irradiation. This synergistic effect was not observed when the ultrasound (cw 1MHz, 2W/cm^2 for 30 min) was given either before the γ rays or more than 2 h afterwards.

In another study, the exposure of tumour cells to ultrasound and X-rays reduced the electrophoretic mobility of the cells by 30% (Repacholi, 1970). The author proposed that ultrasound and X-rays might have been capable of shearing the mucopolysaccharide coat from the tumour cell, thus enhancing the potential for tumour-cell killing by lymphocytes.

6.3.4 Summary

Ultrasound exposure apparently alters both cellular ultrastructure and metabolism. Cells exposed to ultrasound appear to be more prone to cell death during mitosis. Supression of cellular growth has been reported under cw and pulsed exposure conditions. Cellular and molecular effects of ultrasound at low SATA intensities are given in Table 11, where many of the effects have resulted from pulsed exposures. This, of course, could be at least partially due to other non-acoustic factors, where, for example at studies in which these effects were observed involved more sensitive end-points.

Table 11. Cellular and molecular level effects

SATA intensity (mW/cm²)		Total exposure time (min)	Effect observed	Reference
less than 0.1	(p)	7.5 to 90	increased rate of sister chromatid exchange (lymphocytes)	Haupt et al. (1981)
0.9	(p)	0.5	attachment of cultured human cells	Siegel et al. (1979)
2.5	(p)	0.3	disorganization of microtubules	Cachon et al. (1981).
2.61	(p)	30	alterations of electro-kinetic potential and erythrocyte agglutination	Hrazdira & Adler (1980)
2.7 and 5.0	(p)	30	increased rate of sister chromatid exchange (lymphocytes)	Liebeskind et al. (1979b)
10	(cw)	30 and 90	no change in rate of sister chromatid exchange (Chinese hamster ovary cells)	Wegner et al. (1980)
15	(p)	up to 40	unscheduled non-S-phase (repair) DNA synthesis	Liebeskind et al. (1979a)
15	(p)	up to 40	disturbances in cellular growth pattern	Liebeskind et al. (1979a)
15	(p)	30	ultrastructural changes (mouse fibroblasts and rat peritoneal cells)	Liebeskind et al. (1981a)
15	(p)	30	changes in topography of cell surface	Liebeskind et al. (1981a)
15	(p)	30	hereditary changes in cell mobility (mouse fibroblasts)	Liebeskind et al. (1981b)
20	(cw)	10	increase in chromosomal aberrations when given before X-ray exposure	Kunze-Muhl (1981)
40	(cw)	3	altered visco elastic properties (Elodea cells)	Johnson & Lindvall (1969)
60	(p)	30	suppression of cell growth	Maeda & Tsuzaki (1981)

Table 11 (contd).

SATA intensity (mW/cm^2)		Total exposure time (min)	Effect observed	Reference
200	(cw)	15	damage to DNA (calf thymus)	Galperin-Lemaitre et al. (1975)
200	(cw)	5	increase in protein synthesis (hepatic, renal, and myocardial tissue)	Belewa-Staikowa & Kraschkowa (1967)
250	(cw)	0.5	changes in topography of cell surface (m3-1 cells)	Martins (1971)
400	(cw)	3	degradation of DNA (calf thymus and salmon sperm)	Hill et al. (1969)
500	(cw)	10	changes in protein metabolism	Bernat et al. (1966a)
500	(cw)	5	ultrastructural changes (human fibroblasts)	Harvey et al. (1975)
500	(p)	5	ultrastructural changes (human fibroblasts)	Harvey et al. (1975)
500	(cw)	5	increase in permeability of human erythrocyte membranes to potassium ions	Lota & Darling (1955)
600	(cw)	30	decrease in transport of leucine in avian erythrocytes	Bundy et al. (1978)
800	(cw)	60	suppression of cell growth	Maeda & Murao (1977)
900	(cw)	0.3	ultrastructural changes (HeLa cells)	Watmough et al. (1977)
1000	(cw)	5	retarded protein synthesis	Belewa-Staikowa & Kraschowa (1967)
3000	(cw)	10	increase in chromosomal aberrations when given after X-ray exposure	Kunze-Muhl (1981)
36 000	(cw)	10	no sister chromatid exchanges	Morris et al. (1978)

6.4   Effects on Multicellular Organisms

6.4.1 Effects on development

To date, most of the work on the effects of ultrasound on development has been carried out on Drosophila melanogaster, the mouse, and the rat.

6.4.1.1 Drosophila melanogaster

Many studies have been performed on the eggs, larvae, and prepupal stages of Drosophila melanogaster and a variety of abnormal developmental effects have been observed in the adult flies (Fritz-Niggli & Boni, 1950; Selman & Counce, 1953; Child et al., 1981a, b). With the possible exception of eggs in the early stages of development, all insects contain microscopic, stable gas bodies throughout their life cycle. These gas bodies oscillate under the influence of the ultrasound and presumably generate streaming motions in adjacent soft tissues, that are probably responsible for the observed effects. The results of these studies may not be applicable to mammalian systems, which apparently do not contain stable gas bodies of comparable dimensions.

6.4.1.2 Mouse

Much of the work conducted on developmental effects in mice has been concerned with the use of very high ultrasound intensities and the observed effects were most probably due to heating. Such studies are of very limited value for a health risk assessment from ultrasound exposure and have therefore not been included.

Early mouse morulae (2-4 cell embryos) were exposed to focused and pulsed diagnostic ultrasound in vitro (2.25 MHz, 2.2 mW/cm^2, repetition rate 500 Hz, pulse duration 3 µs) for 12 h; no suppression of growth was observed (Akamatsu & Sekiba, 1977). Hara et al. (1977) exposed 8-day-old mouse embryos to pulsed ultrasound (2 MHz, pulse duration 180 µs, repetition rate 150 Hz) for 5 min. The animals received SATA intensities of either 50 mW/cm^2 or 600 mW/cm^2; an increased incidence of fetal malformations was observed following the higher intensity exposure. At this higher intensity (SPTP intensity 22 W/cm^2), a temperature rise of about 3 °C was measured. The authors also reported a signif- icant reduction in maternal weight following exposure to ultrasound.

When 8-day-old mouse embryos were exposed to ultrasound in utero (cw 1 MHz, SATA intensities 0.5-5.5 W/cm^2, 10-300 s), a statistically significant reduction in fetal weight was observed (O'Brien, 1976). This observation was confirmed by Stolzenberg et al. (1980a) using cw 2 MHz ultrasound at SATA intensities of 0.5 and 1 W/cm^2 for 1-3 min. Threshold conditions reported to produce a decrease in the mean uterine weight in the progeny were 0.5 W/cm^2 for 140 s or 1 W/cm^2 for 60 s (Stoltzenberg et al., 1980b). However, temperature measurements showed that the uterine temperature was elevated to more than 44 °C, indicating that damage was due to a thermal mechanism. In these studies, hind-limb paralysis and distended bladder syndrome were observed in the mothers at laparotomy and this may have been a contributing factor in the reported weight loss in the mothers and offspring (Stolzenberg et al., 1980c). Reduced fetal body weight has also been reported by Tachibana et al. (1977) following exposure to cw 2.3 MHz ultrasound at SATA intensities of 80-100 mW/cm^2, and by Stratmeyer et al. (1979, 1981a), who used cw 1 MHz ultrasound for 2 min at a SATA intensities of 75-750 mW/cm^2. Growth-inhibiting effects on fetuses were reported by Shoji et al. (1975) in one of two strains of mice following a 5-h exposure to cw 2.25 MHz ultrasound at an intensity of 40 mW/cm^2. However, Edmonds (1980) contends that the calculated free-field intensity for these experiments was closer to 280 mW/cm^2.

An increased incidence in fetal abnormalities was observed after a 5-min exposure in utero to cw ultrasound of approximately 2 MHz, at a SATA intensity of 1.4 W/cm^2, but not at SATA intensities of 0.5 or 0.75 W/cm^2 (Shimizu, 1977). Hara (1980) also found fetal malformations after an in utero exposure to cw 2 MHz ultrasound at 2 W/cm^2 for 5 min; the uterine temperature rose to 41.5 °C. Similar results were obtained using pulsed, 2 MHz ultrasound (SATA intensity 296 mW/cm^2, pulse duration 5 μs, repetition rate 1 kHz, SATP intensity 59.4 W/cm^2), but not at lower SATA intensities or shorter pulses (Takabayashi et al., 1980). A significant increase in skeletal abnormalities was observed in two strains of mice subjected to the same ultrasonic exposure (cw 2.25 MHz, SATA intensity 40 mW/cm^2, for 5 h), but visible mal- formations were only present in one of the strains (Shimizu & Shoji, 1973).

Curto (1975) observed an increased mortality rate in the mouse offspring exposed in utero to cw 1 MHz ultrasound at SATA intensities of 0.125, 0.25, and 0.5 W/cm^2, for 3 min.

However, Edmonds et al. (1979) did not find any effects on neonatal mortality after exposure to cw 2 MHz ultrasound at a SATA intensity of 0.44 W/cm², for a similar exposure time but at a different gestational age.

6.4.1.3 Rat

The development of pre-implantation morulae and early blastocysts of rat was suppressed after exposure to cw 2 MHz ultrasound at 1 W/cm², and necrotic changes occurred after exposure at 3 W/cm² (Akamatsu et al., 1977). Suppressed development was also noted in early embryos after exposure to pulsed 2 MHz ultrasound (10 μs, SATA intensity 0.6 W/cm², SPTP intensity 220 W/cm²), however, development progressed normally after exposure to an SATA intensity of 20 mW/cm² (Akamatsu, 1981).

An extrapolated threshold intensity of about 3 W/cm² was found to be lethal for rat fetuses in utero, subjected to cw 0.71 or 3.2 MHz ultrasound for 5 min (Sikov et al., 1976). The susceptibility of the fetuses depended on the gestational age at the time of exposure. Increased fetal anomalies without corresponding decreases in fetal weights were reported by Sekiba et al. (1980) following exposure to cw 2 MHz ultrasound (SATA intensities 1.5 and 2.5 W/cm²) for 15 min. In a study by Sikov et al. (1977), rat fetuses were exposed in utero to cw 0.93 MHz ultrasound (SATA intensities of 0.01-1 W/cm²) for 5 min; an increased incidence of prenatal mortality and delayed neuromuscular development were found. However, the authors did not find any evidence of increased postnatal mortality or reduced growth rate. A slight (but not statistically significant) increase in skeletal variations and resorption rates was reported by McClain et al. (1972) following in utero exposure to cw 2.5 MHz ultrasound at an SATA intensity of 10 mW/cm² for 0.5 or 2 h, at various gestational ages. No significant differences were observed in viability, body weight, litter size, implantation, and skeletal or soft tissue abnormalities.

Pulsed ultrasound exposures were reported to have caused an increased incidence of gross and microscopic heart anomalies in rat fetuses exposed to 2.5 MHz at SATA intensities greater than 0.5 W/cm² or SATP intensities greater than 50 W/cm² (Sikov & Hildebrand, 1977). More extensive studies failed to confirm the occurrence of cardiac anomalies but did confirm changes in neuromuscular development at SATA intensities greater than 0.5 W/cm² (Sikov, personal communication). Takeuchi et al. (1966) did not find any significant increase in the number of malformations or any change in fetal weight in rat fetuses exposed in utero to a

pulsed, 1 MHz clinical apparatus. Similar negative results
were reported by Shimizu & Tanaka (1980), who exposed pregnant
Chinese hamsters to pulsed, 2 MHz ultrasound (3-µs pulses, 1
kHz repetition rate, SATA intensity 200 mW/cm², SATP intensity 67 W/cm²) for 5 min on days 8, 9, and 10 of gestation.

6.4.1.4 Frog

Sarvazyan et al. (1980) exposed explants of embryos of
Rana temporaria, at different stages of development, to 1 MHz
ultrasound (SATA intensity 50 mW/cm², pulse repetition
frequencies in the kilohertz range, duty factor, 0.5). Local
necroses and complete blockage of gastrulation, observed after
15 min exposure, were highly dependent on the pulse repetition
frequency. The ultrasound did not seem to be as effective in
inducing effects after gastrulation had occurred.

6.4.1.5 Summary

Reports on the effects of ultrasound on animal development
are summarized in Tables 12 and 13.

Table 12. Weight reduction in mice

SATA intensity (mW/cm²)		Total exposure time (min)	Effect observed	Reference
2000	(cw)	5	reduced maternal weight	Hara et al. (1977, 1980)
1000	(cw, p)	8.8	reduced fetal weight	Stolzenberg et al. (1980a)
500 - 5000	(cw)	0.16-5	reduced fetal weight	O'Brien (1976)
500 - 1000	(cw)	1-3	reduced fetal weight	Stolzenberg et al. (1980b)
80	(cw)	8	reduced fetal weight	Tachibana et al. (1977)
75	(cw)	2	reduced fetal organ weight	Stratmeyer et al. 1979, 1981)
50	(p)[a]	5	reduced maternal weight	Hara et al. (1977)

[a] 22 W/cm² Temporal Peak Intensity.

Table 13. Reports of fetal abnormalities observed in rodents

SATA intensity (mW/cm^2)		Total exposure time (min)	Effects reported	Reference
3000	(cw)	5	fetal abnormalities and prenatal death threshold (rats)	Sikov & Hildebrand (1977)
2000	(cw)	5	increase in fetal malformations (mice)	Hara et al. (1977, 1980)
1400	(cw)	5	fetal abnormalities (mice)	Tachibana et al. (1977)
1400	(cw)	5	fetal abnormalities (mice)	Shimizu (1977)
600	(p)[a]	5	fetal abnormalities (mice)	Hara et al. (1977)
586	(p)[a]	5	fetal abnormalities (mice)	Takabayashi et al. (1980)
500	(p)[b]	5	fetal heart abnormalities (rat)[d]	Sikov & Hildebrand (1977)
296	(p)	5	fetal abnormalities (mice)	Takabayashi et al. (1980)
125	(cw)	3	postpartum mortality (mice)	Curto (1975)
40	(cw)[c]	300	fetal abnormalities (mice)	Shoji et al. (1975)
10	(cw)	30	skeletal variations (rats)[e]	McClain et al. (1972)

[a] 22 W/cm^2 Temporal Peak Intensity.
[b] 50 W/cm^2 Temporal Peak Intensity.
[c] This exposure was in air; the calculated equivalent free field intensity in a water bath has been suggested to be 280 mW/cm^2 by Edmonds (1980).
[d] Not statistically significant and not confirmed in a more extensive study by the same investigators.
[e] Not statistically significant.

These reports are difficult to interpret and, in most cases, to compare directly, partly because of differences in the organism used, the state of fetal development at the time of exposure, and the exposure variables. The published works show that, if the intensity is sufficiently high, death or some type of anatomical abnormality will result in certain organisms. Ultrasound is known to raise the temperature of biological samples by which it is absorbed. The effects of exposure at therapeutic intensities (O'Brien, 1976; Stolzenberg et al., 1978; Torbit et al., 1978) are most likely due to hyperthermia (Lele, 1975). Hyperthermal effects in rats and mice depend on the stage of development and exposure conditions, and include fetal resorption, retardation of growth, exencephaly, and defects of the tail, limbs, toes, and palate.

In Table 12, the lowest levels at which fetal weight reduction occurred are in the range 50-80 mW/cm². Within this intensity range and under the experimental conditions used in these investigations, the effects are less likely to be due to hyperthermia. Furthermore, the results of a study by Sarvazyan et al. (1980) suggest that the biological effects induced by pulsed ultrasound may be critically dependent on the pulse repetition rate as well as on the acoustic intensity.

6.4.2 Immunological effects

Effects of ultrasound on the immune response have not been extensively investigated.

Anderson & Barrett (1979) reported a slight, dose-dependent immunosuppressive effect in mice exposed to 2 MHz ultrasound at a SATA intensity of 8.9 mW/cm² (SPTP intensity 28 W/cm²), applied over the area of the spleen. However, the complexity of this response, and the imprecision of the assay techniques used warrant cautious interpretation of these data. Child et al. (1981c) using a similar exposure regime were unable to confirm the findings of Anderson & Barrett (1979).

Mice sonicated over the liver with pulsed 2 MHz diagnostic ultrasound (pulse repetition rate 691 Hz, exposure time 1.6, 3.3, and 5 min, SATA intensity 8.9 mW/cm²) had an impaired ability to clear injected colloidal carbon from their blood (Anderson & Barrett, 1981). The phagocytic index and clearance half-time were not lower than normal, immediately after treatment, but were lower, 48 or 72 h after sonication. In a similar experimental arrangement, Saad & Williams (1982) found that SATA intensities of cw 1.65 MHz ultrasound greater than 0.7 W/cm² were needed before a reduction in the rate of clearance of colloidal sulfur particles from rat blood could be detected in vivo.

Other evidence of immunological effects have been reported by Kiski et al. (1975), Bekhame (1977), and Koifman et al. (1980). In addition, Pinamonti et al. (1982) observed a loss of erythrocyte surface antigens following exposure to a pulsed 8 MHz ophthalmological ultrasound device at a SATA intensity of 2 mW/cm², for 30 min (pulse repetition rate 744 Hz).

6.4.2.1 Summary

It is extremely difficult to draw any firm conclusions on the effects of ultrasound on immunological response. Both diagnostic and therapeutic levels have been reported to induce effects.

6.4.3 Haematological and vascular effects

6.4.3.1 Platelets

Blood platelets are extremely fragile cells which, if stimulated, aggregate and release substances that initiate the formation of a clot (Williams, 1974; Brown et al., 1975).

(a) In vitro studies

Ultrasound exposure at a frequency of 1 MHz reduces the recalcification time of platelet-rich plasma at intensities as low as 65 mW/cm² (Williams et al., 1976a). In a study by Williams et al. (1976b), subsequent morphological analysis of recalcified clots revealed the presence of platelet debris, indicating that the ultrasound had apparently ruptured a small portion of the platelet population, releasing adenosine diphosphate (ADP) and other aggregating agents into the surrounding plasma. These agents then induced other platelets to release, resulting in a self-perpetuating cycle of platelet aggregation and release.

Numerous in vitro studies have confirmed that the ultrasound-induced mechanism responsible for platelet aggregation is some form of cavitational activity (Williams et al., 1976b, 1978; Chater & Williams, 1977; Miller et al., 1979).

A variety of threshold SATA intensities determined within the range 0.6-1.2 W/cm² were found to be critically dependent on the pretreatment and rate of stirring of the sample during sonication (Williams, 1982a). The lowest thresholds were obtained when stabilized gas bubbles were deliberately introduced prior to exposure. Using this system, Miller et al. (1979) detected platelet damage from cw 2.1 MHz ultrasound at SPTA intensities as low as 32 mW/cm², and also with a commercial cw Doppler device. Using a burst (gated) regime (burst duration 1 ms, duty factor 0.1) reduced this threshold to an SPTA intensity of 6.4 mW/cm² (Miller et al., 1979).

(b) In vivo studies

Little information exists in the literature on the effects of ultrasound on platelets in vivo. Williams (1977) demonstrated that shear stress forces, similar to those that might be generated in vivo by acoustic cavitation, could trigger platelet aggregation and the formation of thrombi within intact blood vessels in mice. Effects ranged from platelet adhesion to the endothelial walls of the blood vessel to clot formation and complete occlusion of the vessel. Zarod & Williams (1977) found small platelet aggregates within the microcirculation of the guinea-pig pinna after in vivo exposure to cw ultrasound of either 0.75 or 3.0 MHz for 2 min, at a SATA intensity of 1 W/cm². Platelets that had been only partially stimulated by ultrasound were less likely to respond to other stimuli, such as ADP, for a period of time (i.e., they had become refractory) (Chater & Williams, 1977). Such an effect has also been reported in vivo by Lunan et al. (1979), who found decreased aggregation of platelets after whole-body exposure of mice to cw 2 MHz ultrasound at a SATA intensity of 1 W/cm².

Plasma levels of beta-thromboglobulin (a human platelet-specific protein) were measured by Williams et al. (1977, 1981) after in vivo exposure to cw 0.75 MHz ultrasound at a SATA intensity of up to 0.5 W/cm², but no changes were detected.

Ultrasound-induced platelet effects could have serious clinical consequences. For example, the production of platelet aggregates in vivo might lead to the blockage of circulation in small capillaries and subsequent complications of embolism and infarction, especially in patients exhibiting clinical conditions that might predispose them to thrombosis (e.g., during pregnancy or after surgery). However, some of these interactions may, in fact, be beneficial. For example, Hustler et al. (1978) found inhibition of experimental bruising in the guinea-pig ear after exposure to 0.75 MHz at 0.6 W/cm².

6.4.3.2 Erythrocytes

(a) In vitro studies

Red blood cells are less sensitive to rupture by shear stress than platelets (Nevaril et al., 1968; Rooney, 1970; Williams et al., 1970; Leverett et al., 1972). Veress & Vincze (1976) reported that haemolysis occurred in vitro at intensities as low as 200 mW/cm² (spatial average). It was not determined whether this represented a threshold value, but a linear relationship existed between the logarithm of the time necessary to produce haemolysis at 1 MHz and the intensity of the ultrasound, at a given concentration of blood cells.

In a study by Koh (1981), the blood of pregnant women was exposed in vitro to cw 20 mW/cm² ultrasound for 2-12 h and 2.6 W/cm² for 40-120 min. An increased free haemoglobin level was reported only after exposure to the higher intensity. Significant lysis of human erythrocytes exposed in vitro for 6-8 h to Doppler ultrasound at intensities in the range of 10-20 mW/cm² was reported by Takemura & Suehara (1977). However, Kurachi et al. (1981) reported that haemolysis of human blood did not increase after in vitro exposure of 24 h to a pulsed diagnostic device or 60 min to pulsed 2 MHz ultrasound at 0.57 W/cm² (10 µs pulses, SATP intensity 50 W/cm², pulse repetition rate 1 kHz).

Functional changes in human erythrocytes have been found after in vitro exposures for 30 min to pulsed 8 MHz ultrasound at 2 mW/cm². Irradiation appears to affect the erythrocyte membrane, causing a decrease in the oxygen affinity of the cells (Pinamonti et al., 1982).

(b) In vivo studies

Williams et al. (1977, 1981) were unable to detect haemolysis in human blood exposed in vivo to unfocused cw 0.75 MHz ultrasound at a SATA intensity of 0.34-0.5 W/cm², for an exposure time of about 30 s. However, Wong & Watmough (1980) reported lysis of mouse erythrocytes in vivo after they had irradiated the heart with 0.75 MHz ultrasound at about 0.8 W/cm². This result is probably a reflection of the enhanced nucleation conditions existing within the beating heart. Similar positive results in vivo were reported by Yaroniene (1978), who exposed rabbit hearts to 2 MHz ultrasound in both the cw (SATA intensity 10 mW/cm²) and pulsed modes (pulse duration 4 µs, repetition rate 1 kHz, SPTP intensity 90 mW/cm², SATA intensity 0.4 mW/cm²) for prolonged exposures of up to one month.

6.4.3.3. Blood flow effects

An ultrasonic standing wave field can stop the flow of blood cells within intact blood vessels in vivo (Schmitz, 1950; Dyson et al., 1971; ter Haar, 1977). This effect was subsequently called "blood stasis" or "blood flow stasis" (Dyson et al., 1971). Dyson & Pond (1973) and Dyson et al. (1974) found that the blood cells grouped into bands, spaced at half-wavelength intervals and separated by regions of clear plasma. The bands were oriented in a direction perpendicular to that of the propagating ultrasound. At 3 MHz and high intensities, the minimum time for banding to occur in front of a perfect reflector was approximately 0.05 s. The minimum intensity required for stasis was generally less than 0.5 W/cm^2 at 3 MHz and varied with the type, size, and orientation of blood vessels and with the animal's heart rate. Electron microscopic examination revealed damage to some of the endothelial cells lining the blood vessels in which stasis had occurred. With short exposure times, the effect and damage generally appeared to be reversible. Permanent damage was observed following an extended exposure time of 15 min.

Blood flow stasis has also been observed in mouse uterine blood vessels (ter Haar, 1977; ter Haar et al., 1979). The mechanism responsible for this phenomenon is the radiation force associated with the standing wave field (ter Haar & Wyard, 1978). The authors observed that blood stasis did not occur when the transducer was moved over the irradiated tissue. This is of obvious significance in the therapeutic use of ultrasound where it is normal practice to keep the transducer in motion during treatment.

6.4.3.4. Biochemical effects

Various biochemical alterations have been reported following in vivo exposure of guinea-pigs (Straburzynski et al., 1965; Bernat et al., 1966a) and rats (Sterewa, 1977) to therapeutic levels of ultrasound. Glick et al. (1981) reported chemical and haematological changes in the blood of mice following ultrasonic exposure.

6.4.3.5 Effects on the haematopoietic system

Haemorrhaging was observed in the bone marrow of canine femurs exposed to 500 mW/cm^2 for 2 min (Bender et al., 1954).

Damage to the bone marrow was also observed by Payton et al.
(1975), who exposed dog femurs to cw 875 kHz ultrasound at a
SATA intensity of 2.5 W/cm², for 5 min each day, for 10 days
over a 14-day period, using a slow stroking technique.
Exposure for 5 min to 2.5 W/cm² resulted in a 5 °C increase
in the temperature of the bone marrow cavity. Using the same
technique, a 10-min exposure resulted in gross changes,
including an increased peripheral blood clotting time.

6.4.3.6 Summary

Some of the reported effects of ultrasound on blood are
summarized in Table 14. Strong standing wave fields can stop
the flow of blood in small blood vessels. Prolonged stasis may
cause irreversible endothelial and blood cell damage and the

Table 14. Effects of ultrasound on the blood

Ultrasound intensity	Total exposure time	Effect observed	Reference
4 W/cm² (cw)	10 min	decreased glutathione level and increased ascorbic acid level (guinea-pig, in vivo)	Straburzynski et al. (1965)
65 mW/cm² (cw)	5 min	decrease in clotting time (human blood, in vitro)	Williams et al. (1976a, 1976b)
32-64 mW/cm² (cw) SPTP	1 & 10 min	clumping in platelet rich plasma (human, blood) in vitro)	Miller et al. (1978)
6.4-12.5 (p) mW/cm² SPTA	1 & 10 min	clumping in platelet rich plasma (human blood, in vitro)	Miller et al. (1978)
1 W/cm² (cw)	200 s	biochemical and haematological changes (mouse, in vivo)	Glick et al. (1981)
2 mW/cm² (p)	30 min	Functional changes in erythrocytes (human, in vitro)	Pinamonti et al. (1982)

initiation of blood coagulation. Blood cells in suspension in vitro are lysed at therapeutic intensities (around 1 W/cm²) and at lower intensities if the cell suspensions are stirred or agitated or if gas bubbles are deliberately introduced into the medium. Some functional effects on blood cells have been reported at diagnostic intensities, but these have not been independently confirmed and the mechanism of interaction that produces these effects is not known.

6.4.4 Genetic effects

This section will cover the effects of ultrasound on chromosome aberrations, mutagenesis, and other indicators of genetic damage. For the purpose of this review, genetic effects will include heritable effects or indications of DNA damage in somatic cells as well as genetic cells.

6.4.4.1 Chromosome aberrations

A number of early studies (for review see Thacker, 1973) revealed that exposure to ultrasound induced chromosome aberrations in plant root tips. In most studies, the damage was thought to be a result of cavitation or heating. However, Slotova et al. (1967) reported chromosome aberrations in Vicia faba root tips exposed to ultrasound intensities of 200-300 mW/cm² for 1-20 min, with the number of aberrations returning to normal levels 24 h after irradiation. Gregory et al. (1974) and Cataldo et al. (1973), using intensities of 1-20 W/cm² for up to 2 min, did not observe any "classical" chromosome aberrations in Vicia faba root tips. They did, however, report the appearance of bridged and agglomerated chromosomes in the exposed cells, but not in the control cells. The authors suggested that the standard chromosome aberrations scoring technique would not be suitable for the type of damage seen in these studies, because the "standard" technique is to choose only well-spread metaphase chromosomes for scoring. The significance of the bridged and agglomerated chromosomes is not known.

In the early 1970s, a number of studies were carried out on chromosome aberrations in human and other mammalian cells after ultrasound irradiation. These studies were stimulated, at least in part, by Macintosh & Davey (1970, 1972), who reported the production of chromosome aberrations in human lymphocytes. However, in other studies, which covered a range of variables (frequency, intensity, duration of exposure, cell

7

stage), there was not any evidence of chromosome aberrations after ultrasound exposure (Boyd et al., 1971; Buckton & Baker, 1972; Hill et al., 1972; Watts et al., 1972; Rott & Soldner, 1973). Two studies (Watts & Stewart, 1972; Galperin-Lemaitre et al., 1973) in which cells were exposed in vivo also failed to show chromosome aberrations. Furthermore, when Macintosh et al. (1975) tried to reproduce their earlier work as closely as possible, they were unsuccessful. The preponderance of evidence suggests that diagnostic levels of ultrasound do not cause chromosome aberrations in mammalian cells, but this does not negate the possibility of other genetic damage.

6.4.4.2 Mutagenesis

Thacker (1974) used the yeast Saccharomyces cerivisiae to test for the genetic effects of ultrasound. Two of the assays tested for mutations in nuclear genes, one for mutation in mitochondrial DNA, and one for recombination of a nuclear gene. The exposure variables were similar to those used in diagnostic ultrasound (peak intensity of 10 W/cm^2, using 20-μs pulses and a duty factor of 0.004) or therapeutic ultrasound (cw 5 W/cm^2, for up to 30 min). Tests were also made under more severe conditions than those found in medical applications. None of these exposures showed any evidence of increased mutations or recombination after ultrasound exposure, except under conditions where heat or hydrogen peroxide was allowed to accumulate.

In another mutation study, Thacker & Baker (1976) tested for evidence of mutation in Drosophila melanogaster after exposure to diagnostic levels of ultrasound. There was no evidence of lethal recessive mutations or non-disjunction with ultrasound intensities up to 2 W/cm^2, even though these levels were high enough to kill considerable numbers of flies.

Bacteria have also been used to test for mutation induction after exposure to ultrasound. Combes (1975) used Bacillus subtilis to test for reversion of an auxotrophic mutant after ultrasound exposure. No mutants were seen in this system after exposure to pulsed, 2 MHz ultrasound at intensities of up to 60 W/cm^2.

Genetic damage was studied in mice, in which the gonads had been exposed to cw or pulsed 1.5 MHz ultrasound at 1 W/cm^2 (Lyon & Simpson, 1974). The authors tested for induction of translocations of chromosome fragments in spermatocytes and for the induction of dominant lethal mutations in females. The tests were negative, but because of sample variation and the small number of animals used, only pronounced mutagenic effects would have been observed.

Liebeskind et al. (1979a) found that ultrasound affected
several test systems in cultured mammalian cells, suggesting
possible genetic damage. A diagnostic ultrasound device was
used, and cells were exposed to pulsed 2.5 MHz ultrasound for
20-30 min at a SPTP intensity of 35.4 W/cm². One test system
involved antinucleoside antibodies, which are specific for
single-stranded or denatured DNA, are normally bound only
during the DNA synthesis or S-phase, and have low binding
during the G-1 phase. After ultrasound exposure, the cells
showed increased binding during the G-1 phase, though there
was no evidence of strand breakage as indicated by
alkaline-sucrose gradient ultracentrifugation.

Another test system used in this study was the
incorporation of ³H-thymidine into non-S-phase cells as a
measure of repair synthesis. Exposure to ultrasound resulted
in an increased labelling in the non-S-phase cells, suggesting
an increase in repair synthesis. There was, however, no
evidence of an increase in SCE in HeLa cells (section
6.3.1.2). In the same study, Liebeskind et al. (1979a)
investigated the effects of ultrasound exposure on the
morphological transformation of 10T-1/2 cells and found that
it resulted in the induction of type II morphological
transformants, both with and without the promoter TPA.

In a subsequent study, Liebeskind et al. (1979b) reported
that diagnostic levels of pulsed 2.25 MHz ultrasound induced
small, but significant, increases in SCE in fresh human
lymphocytes as well as in a human lymphoblast line. The
significance of SCE is unknown, but it does appear to reflect
chromosome damage. The increased SCEs reported in this paper
following exposure to high SPTP, low SATA intensities of
pulsed ultrasound are consistent with the findings of Haupt et
al. (1981) but contrary to the findings of Morris et al.
(1978) and Wegner et al. (1980), who used cw exposure
conditions. Morris et al. (1978) exposed human leukocytes to
cw 1 MHz ultrasound at intensities of 15.3-36 W/cm² for 10
min. No increase in SCE was observed after exposure.

Hereditary changes were observed in cell surface characteristics (persisting for 50 generations in culture) and cell
mobility (persisting for 10 generations after a single
exposure to ultrasound) (Liebeskind et al., 1981 a, b). Moreover, changes in cell growth regulation (transformation
assays) suggest that genetic damage does occur after in vitro
exposure of cell suspensions to pulsed diagnostic ultrasound.
It is not clear how these results can be interpreted in terms
of in vivo exposure or extrapolated to human exposure. The
observed immunoreactivity suggests disturbances in cellular
DNA, but other interpretations are possible. The density
gradient analysis does not appear to indicate DNA strand
breakage, but the transformation data suggest possible genetic
damage.

Three types of abnormal morphology of transformed cells
have been described (Reznikoff et al., 1973). The cells used
in this study initially had type I morphology and the
ultrasound treatment transformed a few of the colonies to type
II morphology. Because transformation does not appear to be a
sudden event, but rather a progression of changes (or stages),
and because the transformation seen in this study is
apparently only a part of that progression, it does not
necessarily follow that genetic damage has occurred. It is
significant, however, that ultrasound had an effect on the
process of transformation.

Fahim et al. (1975, 1977) claimed that testicular
sterilization could be achieved in rats by an ultrasound
exposure of 1-2 W/cm² (apparently at 1.1 MHz) and that, from
the evidence of parallel experiments with heating applied by
other means, the ultrasonic action was not purely thermal in
nature. These authors further reported that there were no
genetic abnormalities in the progeny of treated animals in
which reduced fertility was observed.

6.4.4.3 Summary

It is not known if ultrasound, under the exposure
conditions used in diagnostics or therapy, can induce genetic
effects. Hereditary changes have been observed in cells
exposed to diagnostic intensities in vitro and, though the
results cannot be extrapolated to the in vivo situation, they
do suggest the need for further in vivo investigations.

At present, there seems to be little evidence that
ultrasound produces mutations or chromosomal aberrations in
mammalian cells. The best evidence of a possible genetic
effect is presented by the transformation and SCE data, which
do not by themselves prove genetic damage, but suggest it. The
possible role of cavitation in producing effects in cell
suspension systems and the relevance of cavitation under in
vivo conditions must also be considered.

6.4.5 Effects on the central nervous system and sensory organs

6.4.5.1. Morphological effects

While large numbers of studies have reported the
production of lesions in the central nervous system (CNS)
following exposure to short pulses of very high intensity
focused ultrasound, most were considered inappropriate for
determining health risk assessment and have therefore been
omitted.

Borrelli et al. (1981) reported altered morphology of the synapses following exposure of cat brain to pulsed 1 MHz ultrasound at an SPTP intensity of 300 W/cm² for 0.5-3 s. The authors suggested that the morphological changes in the synapses might explain the irreversible interruption in CNS function. They also suggested that the synapses may be more sensitive to ultrasonic exposure than mitochondria, which have previously been thought to be among the structures most sensitive to damage by ultrasound.

6.4.5.2 Functional effects

Hu & Ulrich (1976) exposed the brains of squirrel monkeys to 2.5-5 MHz ultrasound at intensities ranging from 3 mW/cm² to 0.9 W/cm², and recorded induced potentials using electro-encephalograph (EEG) electrodes that had been implanted within the brain for long periods. The monkeys were found to adapt to the exposure within 3 min in that the evoked potentials disappeared, even though the cw or pulsed sonication was maintained. Amin et al. (1981), in an investigation similar to that of Hu & Ulrich (1976), did not observe any effect on the mammalian EEG during exposure to pulsed ultrasound. They suggested that one possibility for the differences was that the 17 Hz and 35 Hz spectral lines observed by Hu & Ulrich were harmonics of the signal. However, this explanation raises a question as to why other harmonics were not also seen. In addition, it would not explain why the potentials detected by Hu & Ulrich disappeared after 2-3 min of exposure, though the ultrasound exposure continued.

Changes in microphonic potentials of cats' ears were reported following irradiation of the labyrinth of the inner ear through the round window of their ears, with 3 MHz ultrasound (200 and 600 mW/cm² for 1-5 min) (Molinari, 1968a). Molinari (1968b) also noted that these effects were reversible at the lower intensity but were irreversible at the higher intensity, since damage to the neuroepithelium of the organ of Corti had occurred.

In studies by Farmer (1968), the conduction velocity of human axons increased following a 5-min exposure to cw 870 kHz ultrasound at a SATA intensity of either 0.5 or 3 W/cm², but decreased at a SATA intensity of 1-2 W/cm². The low intensity result (0.5 W/cm²) was confirmed by Esmat (1975), but he was unable to confirm the findings at the higher intensities. He proposed that the observed changes resulted from temperature elevation. Using pain sensation in the human hand and arm as an end-point, Gavrilov et al. (1976, 1977) found a wide range of intensity thresholds, depending on frequency (0.9-2.7 MHz) and pulse duration (1-100 μs).

Stolzenberg et al. (1980c) reported hindleg dysfunction and distended bladder syndrome following exposure of pregnant mice to cw 2 MHz ultrasound at a SATA intensity of 1 W/cm^2 for 80-200 s. This demonstrated that both the autonomic and somatic nervous systems were damaged, indicating that prudence is necessary in choosing the site of application and duration of therapeutic ultrasound treatment. Another reported functional change in the mammalian CNS is the reversible suppression of nerve potentials (Fry et al., 1958).

6.4.5.3. Auditory sensations

Gavrilov et al. (1975) noted that pulses of focused ultrasound stimulated the auditory receptors of the labyrinth of a frog. They detected bioelectric potentials in the auditory part of the mid-brain resembling those induced by audible stimuli. Irradiation of the cochlea of human volunteers with 2 MHz focused ultrasound (SPTP intensities 50-200 W/cm^2, pulse duration 1 μs) induced click type auditory sensations. The subjects apparently experienced a hearing sensation similar to that found in subjects exposed to pulsed microwave radiation at power densities of approximately 1 mW/cm^2. In this case, the auditory sensations or clicks had been shown to be due to very localized, minute temperature increases. A similar indirect mechanism could exist for ultrasound, or the effect may be due to a direct response to the pulse pressure.

6.4.5.4. Mammalian behaviour

Abnormal behavioural effects in adults may often be caused by damage to the CNS at an early stage of development in utero. Physically restrained pregnant rats were exposed to cw 2.3 MHz ultrasound at a SATA intensity of 20 mW/cm^2 for 5 h on the 9th day of gestation, and their progeny investigated immediately after birth and 100 days later (Murai et al., 1975a, b). A delay in maturation of the grasp reflex was observed (Murai et al., 1975a). Murai et al. (1975b) tested the same animals at 120 days of age and found that vocalization to handling and escape response from electric foot shock (emotional behaviour) were significantly increased in exposed versus sham and untreated control animals. It was concluded that the emotional behaviour of rats could be influenced by prenatal exposure to ultrasound intensities as low as 20 mW/cm^2.

Altered postnatal behavioural changes were also reported by Sikov et al. (1977a), who exposed pregnant rats to cw 0.93 MHz ultrasound at SATA intensities of 10-100 mW/cm², for 5 min, on the 15th day of gestation. Similar behavioural abnormalities were reported for the righting reflex, head lift, and holding responses. The authors concluded that the threshold for these postnatal effects must be less than 10 mW/cm². However, it was observed that these abnormalities were only transient delays in maturation, relative to normal controls. Brown et al. (1979, 1981) have not been able to repeatedly obtain behavioural effects in mice. These data are summarized in Table 15.

Table 15. Behavioural effects in rats and mice

SATA intensity (mW/cm²)	Total exposure time (min)		Effect observed	Reference
20	(cw)	300	delayed neuromotor reflex development (rat)	Murai et al. (1975b)
20	(cw)	300	altered emotional behaviour (rat)	Murai et al. (1975a)
50 - 500	(cw)	2 - 3	variable results (mice)	Brown et al. (1979, 1981)

6.4.5.5 The eye

The lens appears to be the part of the eye that is most susceptible to ultrasound, because it does not have a blood supply to dissipate heat. A temperature rise above a certain threshold in the lens or cornea results in the formation of opaque regions or cataracts. A number of reports (Preisova et al., 1965; Bernat et al., 1966a, b; Gavrilov et al., 1974; Zatulina & Aristarkhova, 1974; Moiseeva & Gavrilov, 1977; Marmur & Plevinskis, 1978) suggest mechanisms whereby ultrasound could induce cataracts.

Preisova et al. (1965) found that cw 800 kHz ultrasound exposure of the eyes of rabbits, for 2 min at SATA intensities greater than 0.5 W/cm², caused significant changes in the temperature of the cornea. Pulsed diagnostic ultrasound lasting up to 8 min caused a very small increase (0.75 °C) in the temperature of the eye. Zatulina & Aristarkhova (1974) also used pulsed ultrasound (880 kHz, pulse duration 10 ms, SATA intensities 0.2-0.4 W/cm²) and observed alterations in the corneal epithelium, which developed at a later date than those resulting from cw exposure at the same frequencies and intensities.

Lizzi et al. (1978a, b) reported that 2 types of cataracts could be induced in the lens of the rabbit eye using high SPTA intensities (200-2000 W/cm²) of focused 9.8 MHz ultrasound. One was a "haze" cataract, discernible only with slit lamp visualization, and the other a totally opaque cataract, occurring after long exposure times (i.e., after more energy had been deposited). Fig. 7 presents the total amount of energy deposited as a function of the length of exposure necessary to produce a minimum detectable haze cataract. Exposures shorter than 0.1 s required a constant energy deposition, whereas longer exposures required increasing energy input. This can be interpreted in terms of a thermal mechanism, whereby heat does not have time to diffuse away from the site of deposition in a time shorter than 0.1 s. With times longer than 0.1 s, more energy has to be supplied to allow for heat diffusion out of the focal volume. The shape of the threshold curve obtained seems to be consistent with that predicted for thermally-mediated damage (Lerner et al., 1973).

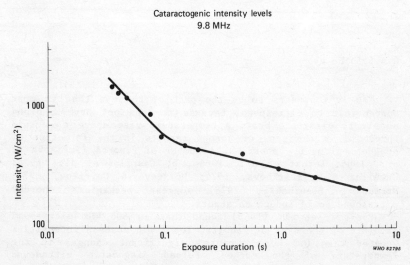

Fig. 7. Total energy-time threshold to produce minimum haze cataracts with 9.8 MHz ultrasound (From: Lizzie et al., 1978b).

Using the same focused experimental system, Lizzi et al.
(1978a) also observed ultrasonically-induced lesions in the
retina, choroid, and sclera. The amount of energy required to
produce a detectable lesion in these parts of the eye was less
than that needed to generate cataracts in the lens or cornea.
Nevertheless, a threshold curve of similar shape was obtained,
which was compatible with the thermal dissipation
characteristics of these structures.

A specialized low-frequency, ultrasonic, surgical
technique (phacoemulsification) has been developed for the
break-up and removal of cataractous lenses. The phacoemulsifier consists of a hollow metal probe oscillating with
displacement amplitudes of the order of tens of micrometres
and frequencies in the range 20-40 kHz. Damage to the
endothelial cells of the cornea has been reported as an
undesirable side-effect of the phacoemulsification procedure
(Talbot et al., 1980). Considerable controversy exists as to
whether or not this damage is the result of ultrasound action
or is the result of other non-acoustic factors associated with
the surgical procedure.

6.4.5.6. Summary

In summary, it can be said that the results of functional
studies are often contradictory, with electrophysiological
measurements showing both increases and decreases. Because of
experimental differences, and dosimetric uncertainties, the
only conclusion that can be reached is that cw power densities
as low as 0.5 W/cm^2 can induce transient alterations in
neural function.

Hindleg paralysis and distended bladder syndrome have been
reported in rodents following exposure to typical therapeutic
intensities of ultrasound. Though the small dimensions of the
rodents would tend to maximize thermal damage, these
observations indicate that the site of application and
duration of exposure of therapeutic ultrasound should be
chosen with care.

Postnatal behavioural effects have been observed in rats
after exposure to 20 mW/cm^2 of cw 2.3 MHz ultrasound as
presented in Table 15. If confirmed, the results of postnatal
functional tests present a serious challenge to the assumption
that fetal exposure to ultrasound is innocuous.

The eye has been identified as an organ sensitive to
ultrasound exposure. Ultrasonically-induced lesions occur in
the retina, choroid and sclera. The lens of the eye is
sensitive to cataract production, probably via a thermal
mechanism.

6.4.6 Skeletal and soft tissue effects

A number of skeletal and soft tissue effects have been reported following exposure to ultrasound. Many investigations have been conducted in this area but, because of the use of ultrasound in physiotherapy, only a few representative examples have been chosen to illustrate the diversity of the observed effects.

6.4.6.1 Bone and skeletal tissue

It has been common practice in physiotherapy to treat the stumps of amputated limbs with high intensities of therapeutic ultrasound, to prevent formation of calcified spur growths from the cut surface of the bone. Unfortunately, there are no known clinical trials to indicate the efficacy of this therapeutic practice, but Kolar et al. (1965) reported that many Eastern European publications have indicated reduced skeletal growth in dogs, after exposure to ultrasound intensities between 3 and 4 W/cm^2. In their own studies, Kolar et al. (1965) used a magnetostrictive ultrasound source (used in dentistry), with an irradiating area of 1.0 cm^2, to deliver static exposure to the knees of young rats for 5 min. A significantly reduced calcium metabolism was observed, at various times, up to 102 days after the exposure, by means of radioisotope tracers.

Barth & Wachsmann (1949) found that young dog bones exposed to ultrasound levels of 0.5-1 W/cm^2 from a stationary transducer showed thickening, followed by loss of the periosteum. Old bones showed similar effects, but they took longer to develop. The authors reported that, for a moving ultrasound field, the threshold limit for bone damage was about 3 W/cm^2.

After fracture of the third metatarsal in rabbits, the fractures were exposed to ultrasound intensities of at least 0.4 W/cm^2. The treatment commenced on the third day, for 8 min daily, with up to 15 treatments. X-ray examinations were used to determine the differences between the control and sonicated group on the tenth day after fracture. Based on histological examination, it was reported that small doses of ultrasound enhanced the process of regeneration, differentiation, and resorption of bone tissue. The fracture was reported to weaken within 10-12 days of cessation of treatment. After 45 days, no differences in the healing of fractures were observed between experimental and control animals (Goldblat, 1969).

6.4.6.2 Tissue regeneration - therapeutic effects

Dyson et al. (1968) reported that tissue regeneration was stimulated by low therapeutic intensities of pulsed and cw ultrasound. They measured the rate of repair of symmetrical 1 cm^2 wounds made in both ears of rabbits. In each animal, the healing process in the wound in the unexposed ear was compared with that in the ear exposed to ultrasound. The 3.6 MHz source used by Dyson et al. (1968) was described by Pond & Dyson (1967). Each treatment involved a 5-min exposure, with 3 treatments given each week. The intensity that stimulated growth was either 100 mW/cm^2 for the cw exposures or in the range 0.25-1 W/cm^2 (peak) for the pulsed exposures (2 ms on and 8 ms off). The observed regeneration rates for the ultrasound-exposed wounds were significantly more rapid than those of the unexposed group. The maximum mean growth increase, which was reported to be about 1.3 times that in the controls, was found 21 days after treatment at 500 mW/cm^2 with a pulse duration of 2 ms and a pulse repetition rate of 100 Hz. The temperature rise resulting from this exposure was 1.5 °C. Because of the low intensity at which this effect was observed and the small temperature rise, it was attributed to a mechanism other than heating (Dyson et al., 1968, 1970; Lehmann & Guy, 1972).

Dyson et al. (1976) also investigated the stimulatory effect of ultrasound in healing varicose ulcers in human subjects. The ultrasound reduced the ulcer area by about 27% compared with untreated controls, 20 days after commencement of treatment. The authors suggested that non-thermal mechanisms might be involved in the action of ultrasound on tissues.

Goralčuk & Košik (1976) reported that when rabbits with Staphylococcus aureus-induced suppurative ulcers of the cornea were treated with ten, 5-min sessions of 1.625 MHz ultrasound at an intensity of 0.4 W/cm^2, plus penicillin, better regeneration of tissue occurred than with penicillin alone. Franklin et al. (1977) irradiated dog hearts, which had myocardial infarcts, with ultrasound (cw 870 kHz, SATA intensity 1 W/cm^2 for 10 min) 3 times a day for 6 weeks. There was less dense collagen scarring in the treated animals, and the infarcted areas, identified by gross and histological examination, were usually smaller in the treated animals.

In general, there are no clinical trials to support the widespread use of ultrasound in physiotherapy (Roman, 1960). However, experienced physiotherapists claim that ultrasound is efficacious in the treatment of many diverse conditions, e.g., in increasing the range of movement at joints. In support of this practice, Gersten (1955) reported increased extensibility

of frog tendon following a 3-min exposure to pulsed 1 MHz
ultrasound (SATA intensity approximately 3 W/cm^2, pulse
duration 1 ms). The higher absorption coefficient of tendon
(collagen) relative to other soft tissues means that this
tissue is selectively heated by ultrasound, which may be the
underlying mechanism responsible for its apparently beneficial
effects (Lehmann & Guy, 1972; Lehmann et al., 1978).

6.4.6.3 Muscle

A change in the spontaneous contractile activity of
mammalian smooth muscle was reported by Talbert (1975),
following exposure to cw 280 kHz ultrasound at an SATA
intensity of 1 W/cm^2, but not following exposure to 2 MHz
ultrasound. Similar contractions, using the same exposure
conditions, have also been found in mouse uterine muscle in
vivo (ter Haar et al., 1978).

Hu et al. (1978) studied the effects of ultrasound on the
smooth muscle of the rat intestine and found that an intensity
of 1.5 W/cm^2 for 5 min at a frequency of 1 MHz inhibited
action potentials. This effect was found to be reversible
following a single exposure, but multiple exposures resulted
in only partial recovery.

When rat cardiac muscle was exposed in vitro to cw 1 MHz
ultrasound (SATA intensity of 2.4 W/cm^2) for 10 min, the
resting tension was altered without a corresponding change in
its active tension (Mortimer et al., 1978).

6.4.6.4 Thyroid

Changes in organ function have been reported for the
thyroid following ultrasound exposures in the therapy range,
i.e., 1 W/cm^2, 0.8 MHz, 10 min (Slawinski, 1965, 1966). Such
exposures were found to result in impaired iodine uptake and,
in animals with marked thyroid hypofunction, reduced
iodothyronine synthesis. Hrazdira & Konecny (1966), who
reported similar findings, indicated that epithelial cells of
the thyroid follicles showed a partial loss in ability to
concentrate inorganic iodine.

Some reports have appeared of whole-body systemic effects
of ultrasonic irradiation, in both experimental animals and
man. Sterewa & Belewa-Staikova (1976) irradiated the lower
abdomen of rats at therapeutic intensities (0.2-1.0 W/cm^2)
and reported a consequent decrease in thyroxin and iodothy-
roxins in the thyroid.

6.4.6.5 Treatment of neoplasia

There has been a revival of interest in the application of
ultrasound for the treatment of malignant tissues. Evidence
has been presented throughout this section that high-intensity
ultrasound, either alone or in combination with other physical
or chemical agents, will kill cells. Earlier work has been
reviewed by Rapacholi (1969) and a comprehensive review of
this topic has also been compiled by Kremkau (1979). Thus only
a brief outline will be presented below.

When solid tumours were exposed in vivo to peak focal
intensities of the order of a kW/cm^2 for short exposure
times, reduced tumour growth rate and volume were observed
(Kishi et al., 1975; Fry et al., 1978). Similar effects have
also been reported following tumour hyperthermia using lower
intensities (0.5-3 W/cm^2, cw) for exposure times of up to 45
min (Longo et al., 1975, 1976; Marmor et al., 1979).

Positive and negative synergistic interactions of
ultrasound and chemicals (Hahn et al., 1975; Heimburger et
al., 1975) or X-rays (Woeber, 1965; Shuba et al., 1976;
Witcofski & Kremkau, 1978) have been reported for the
treatment of cancerous tissues. However, some investigators
have reported conflicting results with different tumour types
treated with the same combination of ultrasonic and X-ray
treatment (Shuba et al., 1976; Witcofski & Kremkau, 1978).

It is not known whether ultrasound could induce metastases
during cancer treatment. However, Siegel et al. (1979), using
diagnostic intensities (approx. 0.62 mW/cm^2), and Ziskin et
al. (1980), using average intensities of between 12 mW/cm^2
and 50 W/cm^2 (880 kHz-2.5 MHz for 5 min-1 h) found increased
cell detachment following exposure to ultrasound in vitro.
Evidence for increased detachment in vivo has not been
obtained, although Smachlo et al. (1979) found that
ultrasonic treatment of hamster tumours (cw 5 MHz, SATA
intensity 3 W/cm^2) for 6-8 min caused a reduction in tumour
growth, and did not cause an increase in the rate of
occurrence of metastases.

6.4.6.6 Summary

The effects of ultrasound exposure on skeletal and soft
tissues are summarized in Table 16. The data seem to indicate
that: (a) damage or retardation of bone growth can occur at
intensities in the range 2.5-4.0 W/cm^2 from a moving transducer, and that damage occurs at lower intensities when the

Table 16. Reported central nervous system, skeletal,
and soft tissue effects

SATA intensity (mW/cm²)	Total exposure time (min)	Effect observed	Reference
1.5 (p)	5	retardation of growth (newt forelimbs)	Pizzarello et al. (1975)
1.5 (p)	360	increased GOT levels in cerebrospinal fluid (canine CNS)	Tsutsumi et al. (1964)
3 (p)	3	evoked transient EEG potentials (primate)	Hu & Ulrich (1976)
8.9 (p)	1.6	effect on liver; depressing phagocytosis (mice)	Anderson & Barrett (1981)
8.9 (p)	5	immunosupressive effect on spleen (mice)	Anderson & Barrett (1979)
10 (cw)	days	microcirculation disturbances (rabbits and frogs)	Yaroniene (1978)
10 (p)	30	fetal skeletal variations (rat)	McClain et al. (1972)
40 (cw)	300	increase in skeletal abnormalities (mice)	Shoji et al. (1971)
50 (p)	15	blockage of gastrulation (frog embryo explants)	Sarvazyan et al. (1980)
80 (cw)	5	stable cavitation (guinea-pig)	ter Haar & Daniels (1981)
100 (cw)	5 (repeated exposure)	wound healing (rabbit)	Dyson et al. (1968)
200	1	reversible changes in evoked microphonic potentials (cat ear)	Molinari (1968a, b)
400 (cw)	10 (repeated exposure)	healing of corneal ulcers (rabbit)	Goralčuk & Košik (1976)

Table 16 (contd).

SATA intensity (mW/cm²)	Total exposure time (min)	Effect observed	Reference
500 (cw)	2	haemorrhaging in bone marrow (dog)	Bender et al. (1954)
500 (cw)	10	change in thyroid function (guinea-pig)	Slawinski (1966)
500 (cw)	-	blood stasis (chick)	Dyson & Pond (1973)
500 (cw)	10	decrease in SH groups (mouse epidermis)	Chorazak & Konecki (1966)
600 (p)	5	fetal skeletal abnormalities (mice)	Hara et al. (1977), Hara, (1980)
500-1000 (cw)	-	bone thickening and loss of periosteum (dog)	Barth & Wachsmann (1949)
1000 (cw)	1.3	hindleg dysfunction (mouse)	Stolzenberg et al. (1980c)
1000 (cw)	1.3	distended bladder (mouse)	Stolzenberg et al. (1980c)
1500 (cw)	-	tissue damage (stationary transducer) (dog)	Hug & Pape (1954)
1000-2000 (cw)	-	tissue damage (stationary transducer) (dog)	Lehmann (1965b)
2000 (cw)	5	fetal skeletal variations (mice)	Hara et al. (1977, 1980)
2400 (cw)	10	change in resting cardiac muscle tension (rat)	Mortimer et al. (1978)
2500 (cw)	10 (repeated exposure)	damage to bone marrow (dog)	Payton et al. (1975)
3000 (cw)	5	bone damage (moving sound field) (dog)	Kolar et al. (1965)
4000 (cw)	-	tissue damage (moving transducer) (dog)	Lehmann (1965b)
300 000 (p) (SPTP)	0.5-3 s	altered synapse morphology (cat)	Borrelli et al. (1981)

transducer is kept stationary; (b) young growing bone appears to be more sensitive to the effects of ultrasound than older bone; (c) tissue regeneration appears to be enhanced by ultrasound exposures at intensities below 2.0 W/cm^2; this seems to be the case for both soft tissue and bone; (d) ultrasound at therapeutic intensities can trigger muscle contractions and inhibit action potentials; (e) ultrasound at therapeutic intensities has also been reported to alter thyroid function; (f) ultrasound alone (hyperthermia) or in combination with other physical or chemical agents may have an application in the treatment of neoplasia.

6.5 Human Fetal Studies

In the quantification of adverse health effects in the fetus, the main problem is the difficulty of demonstrating a causal relationship between exposure to ultrasound and a change in the normal incidence of spontaneous abnormalities. Large groups must be investigated to obtain statistically significant epidemiological data. The problem of adequate control groups is controversial and hinges mainly on what is considered "adequate" (Silverman, 1973).

6.5.1 Fetal abnormalities

There are several frequently quoted studies that claim to show that exposure to ultrasound in utero does not cause any significant abnormalities in the offspring (Bernstein, 1969; Hellman et al., 1970; Falus et al., 1972; Scheidt et al., 1978). However, these studies can be criticized on several grounds, including the lack of a control population and/or inadequate sample size, and exposure after the period of major organogenesis; this invalidates their conclusions as Scheidt et al. (1978) acknowledge.

However, studies incorporating larger sample sizes also do not show any significant differences in the frequency of fetal abnormalities (Morahashi & Iizuka, 1977; Lyons & Coggraves, 1979; Koh, 1981; Mukubo et al., 1981, 1982). Nevertheless, a preliminary analysis of the birth records of 2135 children, exposed to ultrasound in utero, indicated the possibility of fetal weight reduction (Moore et al., 1982). Although the data were adjusted for several confounding factors, not all factors that might affect lower birthweight could be taken into account. While this study does not prove a cause-effect relationship, it does provide guidance for designing further studies.

6.5.2 Fetal movement

David et al. (1975) indicated a significant increase in subjectively assessed fetal activity during routine monitoring of 36 mothers with cw Doppler ultrasound. This result has not been confirmed by either Hertz et al. (1979) or Powell-Phillips & Towell (1979).

6.5.3 Chromosome abnormalities

Several studies have been conducted to determine the incidence of chromosome abnormalities in lymphocytes from fetal and maternal blood exposed to ultrasound in vivo. Only negative or inconclusive results have been reported (Abdulla et al., 1971; Serr et al., 1971; Watts & Stewart, 1972; Ikeuchi et al., 1973).

6.5.4 Summary

There are many gaps in the data from human studies that prevent a meaningful risk assessment of ultrasonic exposure. It is therefore necessary to use the results of animal studies to test the hypothesis that similar effects may also occur in human subjects. Animal studies suggest that neurological, behavioural, developmental, immunological, haematological changes and reduced fetal weight can result from exposure to ultrasound.

Choosing end-points for study is especially difficult in human subjects. Latent periods, before abnormalities become evident, could easily be as long as 20 years, or effects may not be seen for another generation. Many human epidemiological studies have concentrated on the gross developmental abnormalities evident immediately after birth and have yielded negative results with various degrees of statistical confidence. However, a recent human study has indicated a tendency towards reduced birthweight following ultrasonic diagnostic examination during the course of pregnancy (Moore et al., 1982).

It must be realized that not all possible adverse effects have been explored in animal studies and that some potential problems that could occur in man may not be revealed in animal studies. Another difficulty is that the present understanding of the physical mechanisms of interaction of ultrasound with

biological tissue is inadequate and effects obtained following
cw exposures cannot be extrapolated to predict the
consequences of high-peak pulsed exposures at equivalent SATA
intensities (or vice versa).

7. EFFECTS OF AIRBORNE ULTRASOUND ON BIOLOGICAL SYSTEMS

Ultrasound devices are routinely used in a wide variety of industrial processes, including cleaning, drilling, soldering, emulsification, and mixing, as indicated in section 5. Most of these emit airborne ultrasound, not only at the operating frequency, but also at its harmonics. In addition, audible sound is often emitted. Processes such as washing, mixing, and cleaning are generally carried out using high ultrasonic intensities that cause cavitation. This can be seen as a type of boiling in the liquid and is responsible for the emission of high audible noise levels.

The term "ultrasound sickness" (Davis, 1948), which came into use in the 1940s, included such symptoms as nausea, vomiting, excessive fatigue, headache, and disturbance of neuromuscular coordination. No systematic research into the effects of ultrasound was conducted until the late 1950s (Gorslikov et al., 1965). Since that time a few investigators have studied the effects of airborne ultrasound above 10 kHz. Investigations in the laboratory, and in the industrial and general population environments, have shown that the possible effects of airborne ultrasound can be grouped under four headings: auditory, physiological, heating of skin and tissues, and symptomatic effects.

7.1 Auditory Effects

Since the ear is a sound-sensitive organ, much of the research conducted to date has been based on the likelihood that a physical hazard resulting from airborne ultrasound will involve the ear and may result in a measurable effect on hearing sensitivity. Airborne sound or ultrasound is linked with the human body, through the ear, with an efficiency that is 2 or 3 orders of magnitude greater than that by any other route.

Adverse effects are well documented for exposure to high-intensity audible sound below 8 kHz and can be measured as temporary or permanent threshold shifts (TTS or PTS) in sound perception at specific frequencies and sound pressure levels. There has been a lack of suitable hearing test equipment and of a standard for describing normal hearing above 8 kHz; thus threshold shift evaluation above 10 kHz is questionable. Studies conducted to date have relied on control groups that may not have been properly selected, thereby introducing bias into the studies. In a report published by Northern et al. (1962), normal hearing thresholds were given for frequencies above 8 kHz. While this study involves a small and not very representative sample, it does establish a data base that can be used to evaluate data collected in the future.

Examination of octave band sound pressure levels from ultrasonic equipment in the open industrial environment shows equal and sometimes greater dB values in the audible range than at ultrasonic frequencies. Ultrasonic frequencies alone have been reported to generate audible subharmonics in the ear (Von Gierke, 1950a, b) and have been suggested as the cause of auditory effects (Eldridge, 1950). Threshold shift studies conducted by Parrack (1966), Acton & Carson (1967), Dobroserdov (1967), and Smith (1967) showed mixed results. In studies involving military personnel associated with jet aircraft, Davis (1958) could not show any clear auditory or non-auditory effects. Coles & Knight (1965) and Knight & Coles (1966) showed that exposure to airborne ultrasound reduced hearing sensitivity, but with complete recovery. In a review of work to date, Acton (1973, 1974, 1975) and Acton & Hill (1977) concluded that any hazard to hearing from ultrasound frequencies might be due to the high-frequency audible components that are usually present when airborne ultrasonic fields are encountered.

Studies of industrial workers exposed to levels of low-frequency ultrasound, at approximately 120 dB, failed to reveal either temporary or permanent hearing losses (Acton & Carson, 1967). However, TTS were noted in the hearing acuity of subjects taking part in studies conducted by Parrack (1966). He noted TTS at subharmonics of discrete test frequencies in the range of 17-37 kHz in subjects exposed for approximately 5 min to 150 dB airborne acoustic energy. It has long been assumed by investigators that a TTS is a necessary and sufficient condition (over an extended period of time) for a PTS in hearing to occur.

A literature search and a field study conducted by Michael et al. (1974) is the most comprehensive report published to date on the effects of industrial acoustic radiation above 10 kHz.

7.2 Physiological Changes

In studies involving small animals, mild biological changes have been reported during prolonged exposure to airborne ultrasound with levels in the range of 95-130 dB at frequencies ranging from 10 to 54 kHz (Acton, 1974). In studies in man, Ašbel (1965) reported a drop, and Byalko (1964) an increase, in blood sugar levels in workers exposed generally to airborne ultrasound levels of more than 110 dB (Ašbel, 1965). An electrolyte imbalance in nervous tissues was reported by Angeluscheff (1967), and disturbances of sympaticoadrenal activity by Gerasimova (1976). Early reports (Ašbel, 1965; Angeluscheff, 1967) appear to be supported by more recent data (Gerasimova, 1976), where persons exposed to noise underwent a stress reaction that induced similar effects.

Ahrlin & Ohrstrom (1978) reported physiological (non-auditory) effects on human beings exposed to acoustic energy above 10 kHz.

No significant physiological changes were reported in workers as a result of exposure to 110-115 dB at 20 kHz for 1 h (Grigor'eva, 1966a).

7.3 Heating of Skin

Exposure of mice, rats, and guinea-pigs for about 40 min to airborne ultrasound, at sound pressure levels of 150 dB or more, results in death due to excessive body heating, and exposure to 155-158 dB kills the animals in 10 min (Parrack, 1966). Body heating in these animal species was observed at levels exceeding 144 dB at 18-20 kHz (Allen et al., 1948). With a hairless strain of mice, 155 dB were required to induce the same body-heating (Danner et al., 1954). This result can be explained by the fact that fur has a much greater acoustic absorption coefficient than skin (Parrack, 1966).

In man, exposure to airborne ultrasound at 140-150 dB causes vibration of hairs, particularly in the ear canals or nasal openings, and a simultaneous local warming at these sites (Parrack, 1966). A mild warming of the human body surface may occur at 159 dB and the lethal exposure of man to airborne ultrasound has been calculated to be in excess of 180 dB (Parrack 1966).

7.4 Symptomatic Effects

Some workers exposed to industrial ultrasonic sources such as ultrasonic cleaners and drills complained of fatigue, headache, nausea, tinnitus, and vomiting (Acton & Carson, 1967; Acton, 1973, 1974, 1975). At a sound pressure level of 110 dB, and frequencies of 17.6-20kHz, severe auditory and subjective effects, as mentioned above, as well as an unpleasant sensation of fullness or pressure in the ears were reported by Canadian Forces personnel in the vicinity of ultrasonic cleaning tanks (Crabtree & Forshaw, 1977). The sound pressure levels did not exceed 105 dB at the operator's position (20 kHz one-third octave band) or 95 dB (20 kHz one-third octave band) within 4.5 m of the operator.

Changes in vestibular function were reported by Knight (1968) and Dobroserdov (1967) and may explain the reported feelings of nausea. Possible damage to the vestibular labyrinth is indicated in work by Angeluscheff (1954, 1955, 1967). Many of the reported subjective effects occurred at frequencies below 20 kHz and, in fact, may occur only in individuals to whom these frequencies are audible. Nausea, dizziness, and fatigue may involve an interaction of high-frequency, inaudible sound with cochlear or other inner ear

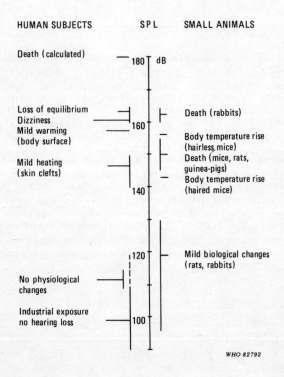

Fig. 8. Reported physiological effects in human beings and animals from exposure to airborne ultrasound (From: Acton, 1974).

functions. Exposure of man to high sound pressure levels of airborne ultrasound causes pressures to be felt in the nasal passage or inside the oral cavity when the mouth is open. Standing wave patterns are frequently set up in these areas (Parrack, 1966).

The audible components of the airborne acoustic energy generated by cavitation in cleaning tanks seem to be directly related to subjective complaints, including fatigue and nausea. However, these complaints may also be attributed to cleaning liquids that have vaporized into the air.

Reports that exposure to airborne ultrasound resulted in neuromuscular incoordination, loss of ability to do mathematical problems, and even complete loss of capacity to perform voluntary acts, appear to be without foundation (Brown, 1967).

7.5 Summary

The physiological effects of exposure to airborne acoustic energy have been summarized in Fig. 8. No adverse physiological or auditory effects appear to occur in man exposed to sound pressure levels up to about 120 dB. At 140 dB, mild heating may be felt in the skin clefts. With increasing sound pressure levels, the human body becomes warmer until death from hyperthermia has been estimated to occur at levels greater than 180 dB.

Subjective or symptomatic complaints such as nausea, vomiting, fatigue, headache, and unpleasant sensations of fullness or pressure in the ears have been reported by persons exposed in the industrial environment. It is difficult to state that the observed effects were due to airborne ultrasound and not audible noise, because many sources of exposure contain acoustic frequencies in both the audible and ultrasonic range.

There is some evidence that any hazard to hearing is probably due to the high-frequency audible sound or to subharmonics of the ultrasonic frequencies. However, it has been reported that temporary threshold shifts in hearing occur after short exposures to airborne ultrasound at 150 dB.

8. HEALTH RISK EVALUATION

8.1 General

At present, there is insufficient clearly established evidence to quantify the health risks resulting from human exposure to ultrasound. In this section, therefore, an attempt is made to put the available scientific evidence into perspective, to identify possible areas of concern, and also to establish criteria that should be satisfied before a meaningful health risk evaluation can be performed.

8.1.1 Criteria

A number of criteria, listed in Table 17, must be considered in a health risk evaluation of the data on biological effects resulting from exposure to ultrasound.

Table 17.   Health risk evaluation criteria for the use of ultrasound: the principles requiring judgement

PRIMARY CRITERIA	WEIGHTING FACTOR
1. Are the data reliable?	Degree of confidence
2. Does the end point relate to a conceivable health risk?	Significance of the health risk
3. Do the exposure-effect data encompass the ranges of human exposure conditions?	Degree of coverage of ranges
4. Can the data be related to in vivo conditions?	Closeness to in vivo conditions
5. Are epidemiological data available?	Statistical significance of findings
6. Is the exposure necessary?	Benefit expected from exposure
7. Are the physical and biological mechanisms understood?	Completeness of understanding
SECONDARY CRITERIA	WEIGHTING FACTOR
a) Is the exposed organism considered to be especially sensitive?	Degree of sensitivity
b) Are the data available from independent sources?	Degree of confirmation
c) Do the data refer to mammalian species	Closeness to human species
d) Exposure condition?	Closeness to exposure condition in human beings
e) Does the exposure occur in combination with other agents?	Extent of interaction

These criteria can be applied to the judgement of a particular publication or to the body of data relating to a particular end-point or biological structure. They are divided into primary criteria, which pose questions of a fundamental nature, and secondary criteria, which are related to the primary criteria and question further details of the studies. Weighting factors are applied to the criteria to provide some quantification and hence to assess the relative significance of the biological effects data for determining health risks.

The following are general examples of how the criteria may be applied to various types of studies to determine their significance for the evaluation of health risk:

i) In vitro studies on molecules in solution showing damage to DNA: though studies of this nature may satisfy certain primary criteria, the data cannot be extrapolated or related to exposure conditions in vivo and such studies cannot be used for health risk evaluation.

ii) In vivo exposure of pregnant mice showing effects on the offspring: this type of study may satisfy the major primary criteria in demonstrating an effect having a significant influence on health risk. If the mechanism is identified as thermal and, as required by the secondary criteria, the data have been independently confirmed, the health risk evaluation revolves around the extrapolation of the ultrasound exposure conditions from the mouse to man. Such an evaluation could take the form of the one performed by Lele (1975).

Obviously, judgements must be made about the usefulness of experimental data in evaluating health risks. Although the criteria show the questions that must be asked, it is the weighting factors that ultimately determine which data indicate the areas of concern. Details relating to these areas of concern in various human exposure situations are discussed in the following section.

8.1.2 Mechanisms

Most of the effects observed in human beings and experimental animals have been attributed to temperature rises resulting from the absorption of the ultrasonic energy by tissues (section 3). Effects expected to follow such temperature rises are the same as those following temperature rises produced by any other agents. Tissue heating is the desired intermediate result in most physiotherapeutic applications of ultrasound. In diagnostic applications, the rate at which energy is delivered to the tissue is usually too low to produce significant heating. During certain types of occupational exposure, tissue heating could occur in combination with other stresses.

Most of the effects observed when using cells in suspension have been attributed to cavitational activity. Cells suspended in a non-absorbing medium are unlikely to be thermally changed, because the absorbed acoustic energy, which is converted into heat, rapidly diffuses out of the cell (Love & Kremkau, 1980). Conversely, individual cells within tissues all absorb the same amount of heat from the acoustic beam, but since there is little net transfer of heat out of the cell, a rise in temperature results in the cells as well as in the surrounding tissues. Thus, in vivo exposures tend to maximize thermal effects, whereas the converse applies to in vitro exposures (Williams, 1982b). However, ter Haar & Daniels (1981) demonstrated that stable gas bubbles (indicative of past cavitational activity) were present, in vivo, in mammalian tissues exposed at SATA intensities as low as 80 mW/cm^2 (0.75 MHz). Also there is evidence of ultrasound-induced effects in blood exposed in vivo, which appear to be the result of cavitation (Yaroniene, 1978; Wong & Watmough, 1980). In this case, the exposures were conducted directly over the heart, where turbulent rheological conditions may have enhanced nucleation (Williams, 1982b).

8.1.3 In vitro experimentation

In view of the considerations outlined above, it can be appreciated that it is very difficult to extrapolate from an in vitro to an in vivo exposure situation. In vitro experimentation allows extensive studies to proceed with reasonable economy of resources. The results of in vitro experiments are extremely valuable for indicating potentially sensitive end-points and interaction mechanisms that should be investigated in in vivo studies.

8.2 Diagnostic Ultrasound

Exposure of patients referred for diagnostic ultrasound examinations may occur once (if the outcome is negative), periodically (for follow-up studies) or intensively for an entire day (for fetal monitoring during labour) (section 5.3.1). Non-intensive examinations are usually completed within 15-30 min.

Long-term occupational exposure of ultrasound technologists and sales and service personnel can result from the practice of using their own organs as test objects to verify correct functioning and desired adjustment of diagnostic ultrasound equipment. This practice should be actively discouraged. Phantom objects are available for these purposes.

Occupational exposure of the hands of technologists, while holding the transducer housing when scanning patients, is conceivable but unlikely to be a significant source of risk.

The acoustic fields relevant to diagnostic exposures are cw fields having space averaged intensities of the order of tens of mW/cm^2, or pulsed fields having SATA intensities of the order of a few mW/cm^2 but composed of microsecond pulses, or bursts having SPTP intensities that may reach 10-100 W/cm^2.

A variety of potentially significant biological effects have been demonstrated in cells in suspension (section 6.3.1). These include changes in cell surface properties, alterations in the rate of macromolecular synthesis and perturbations in genetic material. The interpretation of these results in terms of in vivo exposures is very difficult. The same effects may not occur within the intact organism (when it is subjected to similar exposure parameters), because the mechanism of interaction of the cells with the acoustic field may be different for the reasons previously described (section 8.1.2).

A number of reports on small mammals have indicated a decrease in the average fetal weight following in utero exposure to ultrasonic intensities that have generally been above the levels commonly employed in diagnostic investigations. It has been proposed that high acoustic intensities deposit heat in the fetus, causing a rise in temperature which results in the observed effects (Lele, 1975). This temperature elevation appears to be less likely to occur in the human fetus at typical diagnostic intensities, because of its greater size. However, a similar decrease in fetal' weight was observed in the mouse fetus under conditions in which a temperature rise was considered unlikely (Table 12). It is also of interest that a preliminary analysis of the data on human offspring exposed in utero apparently indicates a statistical association between reduced birthweight and ultrasound exposure (Moore et al., 1982). These findings of possible weight reduction deserve further well controlled investigation, both experimentally and epidemiologically.

Unfortunately, the paucity of data from human studies prevents a meaningful risk assessment being made for

diagnostic ultrasound exposures. Results of animal studies
suggest a wide range of potentially significant biological
changes, including neurological, behavioural, developmental,
immunological, and haematological effects. While most of these
in vivo effects are reported to have been produced at
diagnostic intensities, hardly any have been independently
confirmed, and in most cases the experimental procedures can
be criticised on several points. Bearing in mind that not all
possible adverse effects have been explored in animal studies
and that no single effect (with the possible exception of
fetal weight reduction) is known to be especially sensitive to
ultrasonic exposure, it is not even possible to predict which
biological parameters should to be investigated in human
epidemiological studies.

Additional complications in the choice of suitable
end-points for human studies include: (i) the long latent
period before some abnormalities become evident (which could
easily be as long as 20 years in one individual, or perhaps
even extend into the next generation); and (ii) species-
specific effects may occur in man that may not be revealed in
animal studies.

8.3 Therapeutic Ultrasound

Serial exposure of patients normally occurs in a course of
physiotherapeutic treatments, typically of 5-20 min duration,
repeated daily or intermittently for several weeks. The
ultrasound source may be applied directly to the skin, using a
liquid or gel coupling agent, or both the source and limb to
be treated may be immersed in a water-bath. In recommended
practice, the source is moved continuously to distribute the
absorbed ultrasonic energy throughout the tissue (section
5.3.2).

The frequencies used in therapy range from about 1 to 3
MHz and the SATA intensities from about 0.1 to 3 W/cm^2; the
ultrasound is applied either in a continuous mode or in pulses
that are typically 1 ms or more in duration.

Programmes exist for providing training for physiotherapists in the use of ultrasound, and there have not been
any clearly identified instances of harm to patients from
ultrasound applied according to recommended procedures.
However, it is easy to cause thermal injury (if the source is
not moved continuously) when the higher intensities are
applied. It is common practice to determine the operating
intensity by increasing it to a level just below that at which
the patient experiences pain. This obviously presents the
possibility of hazard, if the patient does not possess normal
sensitivity to pain in the region exposed.

Special caution is necessary in physiotherapy when applying ultrasound to:

(a) bone, particularly growing bone in young children, since heating occurs preferentially at the bone surface;

(b) pregnant women in a manner that might lead to exposure of the fetus, because of the possibilities of fetal abnormalities caused by temperature elevation;

(c) the adult or fetal heart, because of the possibilities of enhanced cavitational activity.

Under normal circumstances, occupational exposure of physiotherapists poses little risk. However, undesirable exposure of the fingers is possible from holding a transducer assembly of faulty design or manufacture, or from placing the hands in a water-bath being used to treat a patient's extremities. Some physiotherapists deliberately subject themselves to unnecessary ultrasonic exposure by routinely using a part of their body (usually the palm of their hand) as a biological test object to check that their transducer is emitting ultrasound. This practice ought to be actively discouraged.

Information on exposure conditions that lead to changes in tissues exposed to ultrasound comes partly from medical experience and partly from studies on laboratory animals and other models. Possibilities for both harmful and beneficial effects exist in the intensity range of therapeutic ultrasound. There is little evidence of therepeutic benefit from the use of SATA intensities of less than 0.1 W/cm^2 and there does not seem to be any need to use SATA intensities greater than about 3 W/cm^2. It is difficult to make a clear assessment of risks versus benefits of exposure to therapeutic ultrasound, because very few clinical studies have been conducted to determine the benefits of the various treatments.

8.4 Hyperthermia

Ultrasound hyperthermia in the treatment of tumours is at present only used as an experimental procedure (section 6.4.6.5). Absorption of focused ultrasound raises the local tumour temperature to 42-44 °C, causing tumour cell destruction. The upper part (1-3 W/cm^2, SATA) of the therapeutic intensity range is used, because rapid heating without tissue disruption is desired. Exposure occurs typically in serial treatments of up to one hour's duration. Ultrasound offers the advantage of effective energy localization for treating deep-seated tumours. Certain hazards obviously attend the procedure.

Repeated low-dose hyperthermia can induce thermal tolerance (Gerner et al., 1976) and possibly stimulate the spread of metastases (Dickson & Ellis, 1974). Superficial burns or fat necrosis can result from inadequate control of local temperatures. The lack of thermal sensors in many deep organs precludes the patient from sensing excessive hyperthermia at these sites. The spinal cord and small bowel may be particularly sensitive to heat or a combination of heat and radiation, and damage to them may be catastrophic (Miller et al., 1976b; Merino et al., 1978; Luk et al., 1980). Metabolic and morphological damage to hepatocytes and neurons occurs at 43 °C (Salcman, 1981). Despite these hazards, hyperthermia treatment has been found to be beneficial for some patients who could no longer tolerate or were unresponsive to more conventional forms of cancer therapy.

Exposure-effect data for ultrasound hyperthermia are sparse and subject to considerable uncertainty. Safety is questionable because tumour temperatures must be raised to at least 42.5 °C for efficacy, while in adjacent normal tissue 45 °C must not be exceeded, if damage due to protein denaturation is to be avoided.

Hyperthermia treatment of tumours requires specialized equipment and expertise. The procedure should not be attempted with equipment and facilities intended for physiotherapeutic applications.

8.5 Dental Devices

Exposure to ultrasound from dental devices occurs when patients have their teeth scaled or cleaned, which typically occurs once or twice annually (section 5.3.5). Adverse effects are quite possible when ultrasound devices in dentistry are improperly used. The problem of avoiding such effects should be solved by the application of suitable training and operative techniques rather than by performing a risk-benefit evaluation, as would be the case in the presence of an unavoidable risk.

The best known effect is due to heating. Modern ultrasonic scaling devices have a water spray or mist for cooling the tool tip and tissue interface. Since the water mist produced by the nozzle obscures vision at the site, optimum water flow adjustment is needed (Frost, 1977). Too much water hinders

operation and possibly drives dislodged calculus into the
gingiva; too little leads to tip heating and patient
discomfort. It is also necessary that the instrument tip be
kept in constant motion to avoid unnecessary "spot" heating of
the teeth (Johnson & Wilson, 1957).

Scratching or "gouging" of teeth by ultrasound scalers can
occur if the tool is applied with too much pressure or insufficient water is used to provide good coupling (Johnson &
Wilson 1957; Moskow & Bressman, 1964; Forrest, 1967;
Wilkinson & Maybury, 1973). The level of training and
experience of the operator are significant factors in the type
of results obtained with ultrasonic scaling. For example, at
the beginning of a training course for dental hygienists,
nearly all the surfaces of the artificial teeth scaled showed
"considerable scratching" whether scaled by hand or by the
ultrasound procedure. Towards the end of a 1-year course,
teeth showed little or no evidence of this scratching
(Forrest, 1967).

These devices typically operate at frequencies in the
range of 20-40 kHz and the tip of the workpiece can be driven
with displacement amplitudes as high as 40 µm. The large
impedance mismatch between the metal probe and the cooling
water, and the intermittent nature of the contact between the
probe and the enamel surface of the tooth, ensure that most of
the acoustic energy is reflected back into the transducer.
Nevertheless, a significant amount of acoustic energy is
transmitted through the treated tooth and may be conducted
through the bones of the upper jaw to the inner ear labyrinth
where it may adversely affect the patient's hearing (Möller et
al., 1976). Damage to the hearing of both the patient and the
operator may also result from airborne ultrasonic energy and
from high levels of airborne audible sound generated by the
cavitational activity occurring within the cooling liquid.

Occupational exposure through direct coupling of the
dental hygienist and the applicator is conceivable but of
minor concern because of the design of the applicators.

8.6 Airborne Ultrasound

Exposure occurs in a variety of occupational and domestic
settings, e.g., in the vicinity of ultrasonic equipment for
cleaning, welding, machining, soldering, emulsifying, drying,
guidance of the blind and robots, intruder detection, TV
channel selectors, and animal scarers. Occupational exposure
is likely to be continuous, while domestic exposure is usually
intermittent and infrequent.

The use of experimental animals to test for biological effects has serious drawbacks because, compared with human beings, they have a greater hearing acuity, wider audible frequency range, and a greater surface-area-to-mass ratio combined with a lower total body mass. Most also have fur-covered bodies. Hence, extrapolation of data from airborne ultrasound studies with animals to man cannot seriously be considered, except in the most general concepts.

In human studies, the hearing acuity of the test and control populations must be considered at exposure frequencies where perception may be audible, audible with recruitment, or inaudible. Recruitment means that the sensation of loudness grows more rapidly than normal, as a function of intensity. Environmental, physiological, and psychological factors that may influence the number of effects observed must also be considered.

There is evidence suggesting that a distinction should be made between inaudible airborne acoustic radiation and that containing audible components (section 7.1). In a study in the USSR, Dobroserdov (1967) concluded that the "effect produced by high-frequency sound was more pronounced than that of the ultrasonic waves". Acton & Carson (1967) suggested that "when these effects occur, they are probably caused by high sound levels at the upper audio frequencies present with the ultrasonic noise". Several other papers report data that indicate the importance of audible components (Skillern, 1965; Acton, 1968). It has also been suggested that the ear, when subjected to high levels of inaudible ultrasound, may generate subharmonics within the audible range and that these subharmonics may be related to some of the observed effects (Eldridge, 1950; Von Gierke, 1950a, b). Thus, it is essential to distinguish the presence or absence of audible components in a given exposure to high-frequency airborne acoustic radiation.

8.7 Concluding Remarks

(a) The levels of human exposure to ultrasound occurring in diagnostic, therapeutic, and dental applications and through airborne ultrasound have been indicated in sections 8.2-8.6. The types of potentially adverse health effects likely to require the setting of limits for safe use, or to require priority in any risk-benefit decisions, have been described. The lowest levels of exposure in in vitro, and experimental animal studies that resulted in quantitative or qualitative biological effects have also been provided, together with the results of some preliminary or small-scale epidemiological surveys on patients after exposure to diagnostic ultrasound. For each of these applications, an attempt has been made to evaluate the existing level of safety or risk. The degree of uncertainty in these evaluations (because of the unavoidable limitations of present knowledge) has been indicated.

(b) There are many deficiencies and gaps in the current data base for ultrasound-induced bioeffects. Most of the data apply to mammals other than man, and it is not usually clear how to relate them to human beings. More information is needed: (i) on the relationship between degrees of risk posed by peak intensities compared with average intensities; (ii) the possibility of cumulative effects; and (iii) the possibility of long-term effects. Also, very few of the data, either positive or negative, have been verified by other than the original reporter. Because of the many difficulties associated with work in the area of ultrasound bioeffects, verification of many more of the data is imperative. These deficiencies and gaps must be resolved before adequate quantification of the safe levels of diagnostic exposure and of any risk-exposure relationships, which are likely to exist at higher levels of exposure, can be achieved. Even at the present level of research activity, it will probably be many years before such quantification can become conclusive. In the meantime, actions and recommendations can and should be taken, based on current data. Recommendations or standards could be revised as more data become available.

(c) Current understanding of the mechanisms of interaction of ultrasound with biological tissues suggests that a specific threshold region may exist for each well-defined exposure-response relationship. However, such threshold regions may vary as the values of physical and biological variables change. If a threshold region exists for any defined response, then sub-threshold exposure would not evoke such a response or cause damage, even after numerous irradiations. However, exposure levels exceeding the threshold region must entail a degree of risk. The practical application of the threshold concept is severely limited by the fact that biological conditions within the living body are subject to large intra- and inter-individual variations. Thus, thresholds tend to become undetectable under marginally supra-threshold exposure conditions as a consequence of this biological variation. It is this fact, together with the uncertainty that the thresholds even exist under experimental or clinical conditions, that limits the use of this concept in risk-benefit considerations rather than any conceivable non-existence of thresholds.

(d) With regard to the "Statement on Mammalian in vivo Ultrasonic Biological Effects" of the American Institute of Ultrasound in Medicine (AIUM, 1978a), the view is still held that it is an adequate statement of the absence of independently confirmed, significant biological effects in mammalian tissues, when the indicated values of exposure to cw ultrasound are not exceeded. This statement, together with some of the comments that accompanied the original publication of the statement, is reproduced in Appendix III. It is clear from these that the statement is a generalization of experimental data for in vivo mammalian systems. Furthermore, its scope is limited in that very few systematic studies have been conducted in which mammalian systems have been exposed to repeated short, high-intensity pulses characteristic of pulse-echo techniques used in diagnostic ultrasound exposures. It is intended to be only a statement of current experimental bioeffect knowledge, and not an immediate recommendation for working levels that must not be exceeded in medical practice.

9. PROTECTIVE MEASURES

The diversity and rapid proliferation of applications of devices emitting ultrasound, combined with reports of potentially adverse health effects, make the need for developing appropriate protective measures increasingly important. Such measures can incorporate safety regulations and guidelines, including the development of equipment performance standards and exposure limits. In addition to specific protective measures, education in this area is very important.

9.1 Regulations and Guidelines

An evaluation of the methods for establishing regulations or guidelines is becoming increasingly important in the field of ultrasound. The identification of effects that pose potential health risks and the way in which limits of exposure might be set in standards relative to the biological effects constitute integral parts of this evaluation.

A standard is a general term, incorporating both regulations and guidelines, and is defined as a set of specifications or rules laid down to promote the safety of an individual or group of people. A regulation is normally promulgated under a legal statute and is referred to as a mandatory standard. A guideline does not generally have any legal force and is issued for guidance only - in other words, it is a voluntary standard. Standards can specify limits of exposure and other safety rules for personal protection, and/ or specify details on the performance, construction, design, or functioning of a device, or methods of testing its performance.

The implementation of standards, which limit exposure to ultrasound, is intended to benefit the health of exposed persons and to provide a frame of reference for industry. Such standards may be useful in the following ways (Repacholi & Benwell, 1982):

(a) Their existence serves as a signal to industry and the general population that there is concern about ultrasound exposure and that they should become aware of the potential hazards.

(b) They provide goals to be achieved at the planning stages by manufacturers of devices and by organizations involved in the installation and construction of ultrasound facilities.

(c) Devices or facilities producing ultrasound in excess of the specified levels should be identified and appropriate remedial action taken.

(d) They form the basis for safe working practices to ensure that workers are not exposed to excessive levels of ultrasound.

Standards that relate to performance or performance testing provide manufacturers and users with standardized procedures for comparing different makes and models of equipment intended to be used for the same general purpose.

Safe-use guidelines have a number of advantages over regulations:

(a) they can be introduced more rapidly;

(b) they can be modified quickly, if necessary; and

(c) they can be specified with more flexibility to adjust to changes in technology.

On the other hand, safe-use guidelines have limitations; because they are voluntary, they need not be heeded, though peer pressure to conform follows from professional and public education on the contents of such guidelines.

9.2 Types of Standards for Ultrasound

To protect the general population, patients, and persons occupationally exposed to ultrasound, two types of standard are generally promulgated:

(1) Emission or performance standards, which refer to equipment or devices and may specify emission limits from a device, usually at a specified distance. Detailed specifications on the design, construction, functioning, and performance of the device are usually given to ensure that the emission limits are not exceeded. An example is the 3 W/cm^2 maximum

output intensity permitted by the Canadian Ultrasound
Therapy Device Regulation (Canada, Department of
National Health and Welfare, 1981). The 3 W/cm^2
limit is also proposed in the draft standard of the
International Electrotechnical Commission (1980b).

(2) Exposure standards, which apply to personnel
protection and generally refer to maximum levels that
should not be exceeded in case of whole or partial
body exposure. This type of standard has greater
applicability to ultrasound as used in industry,
where, for example, exposure standards may limit the
intensity of airborne ultrasound in the environment
of the working place.

Other types of standards that require specific labelling
or disclosure of performance data, or that specify methods of
testing performance, also protect patients indirectly.

Standards development should preferably be preceded by the
preparation of, or reference to, a document that summarizes
the experimental data gained from exposure of various
biological systems to ultrasound, the known mechanisms of
interactions of ultrasound with biological systems, and an
assessment of the various national and international
standards. Such a criteria document can form an important
scientific basis for incorporation of recommendations, from
which the need for exposure limits in standards can be
determined and justified.

9.2.1 Standards for devices

9.2.1.1 Diagnostic ultrasound

It has been stated that "with expanding services in
ultrasound diagnosis, the frequency of human exposure is
increasing with the potential that the major part of the
entire population (of some countries) may be exposed" (IRPA,
1977). The US National Center for Devices and Radiological
Health, using available data on the growth rate of sales of
diagnostic ultrasound equipment, forecasts that the majority
of the children born in the USA after the early 1980s could be
exposed to ultrasound in utero (Stewart & Stratmeyer, 1982).

The following are some standards and test methods for
ultrasound that have been developed and reviewed by Repacholi
(1981) and Repacholi & Benwell (1982).

The International Electrotechnical Commission (IEC) is developing standards for ultrasound medical diagnostic equipment (IEC, 1980a, 1982).

The American Institute of Ultrasound in Medicine (AIUM), through its standards committee, has been very active in the diagnostic ultrasound field. The following are examples of diagnostic ultrasound standards that exist or are being developed:

(i) 100 Millimeter Test Object, including standard procedure for its use (AIUM, 1974);

(ii) American Institute of Ultrasound in Medicine standard on presentation and labelling of ultrasound images (AIUM, 1978b);

(iii) Standard specification of echoscope sensitivity and noise level including recommended practice for such measurements (AIUM, 1979);

(iv) American Association of Physicists in Medicine (AAPM) ultrasound instrument quality control procedures (AAPM, 1979);

(v) Recommended nomenclature: physics and engineering (AIUM, 1980);

(vi) Pulse echo ultrasound imaging systems: performance tests and criteria (AAPM, 1980);

(vii) American Institute of Ultrasound in Medicine standard for transducer characterization (AIUM, 1981);

(viii) AIUM-NEMA safety standard for diagnostic ultrasound equipment (AIUM-NEMA, 1981).

The Acoustical Society of America (ASA) and the American National Standards Institute (ANSI), through their working group S3-54, have undertaken to produce a "performance standard for ultrasonic diagnostic equipment in use". The National Bureau of Standards (USA) is developing standards for application in medicine, industry, and research (National Bureau of Standards (USA), 1973), to be used in connexion with measuring power, intensity, and radiation field patterns of ultrasound transducers.

In France, the Union Technique de l'Electricité has produced a standard for therapeutic ultrasound devices (Association française de Normalisation, 1963). A standard for diagnostic ultrasound devices, which includes specifications on construction, labelling, use, and conditions for approval was published in 1982 (Association française de Normalisation, 1982).

The Japanese Standards Association (JSA) has several industrial standards for diagnostic ultrasound devices. These include A-mode (JSA, 1976), manual scanning B-mode (JSA, 1978), fetal Doppler (JIS, 1979), M-mode (JIS, 1980), and general performance standards (JIS, 1981). Besides safety requirements on electrical parameters, construction, design, and testing procedures, there is a recommendation that would limit the SATA intensity for fetal Doppler diagnostic equipment to no more than 10 mW/cm². For manual scanning B-mode devices, the Japanese Standards Association (JSA, 1978) has many of the same requirements as for A-mode devices, except that it recommends that when tested under specified free-field conditions, the ultrasonic intensity should be less than 10 mW/cm² for each probe; while for M-mode units, the SATA intensity as specified should be less than 40 mW/cm² for each probe. It should be noted that, in theory, the SPTA intensity is four times greater than the SATA intensity but, in practice, the former quantity exceeds the latter by a factor of 2 to 6 (section 2.2.1).

9.2.1.2 Therapeutic ultrasound

Ultrasound has been used since the 1930s in physiotherapy. Though the biological mechanisms of ultrasound therapy have not received systematic investigation, many standards have been developed for therapeutic ultrasound devices. For example, there are both French (Association Française de Normalisation, 1963) and Australian standards (Standards Association of Australia, 1969) on ultrasonic therapy equipment, which indicate ultrasonic output tests and techniques of measurement. Both Canada and the USA have published regulations on ultrasound therapy devices under their respective radiation control acts (Canada, Department of National Health and Welfare, 1981; US Food and Drug Administration, 1978). The International Electrotechnical Commission (IEC) is also developing safety requirements for therapy equipment (IEC, 1980b).

Standards incorporating accuracy specifications for the acoustic output power and intensity and for the timer are needed, since these directly affect the amount of exposure received by the patient. The labelling of individual applicators is necessary to prevent transducers from being connected to the wrong generator, and thereby probably causing significant discrepancies between the acoustic output and the dial indication.

9.2.1.3 Other equipment performance standards

Working groups of IEC subcommittees 29D and 62D are considering standards for the use of ultrasound in dentistry.

9.2.2 Exposure standards

Exposure to ultrasound can be either through direct contact, a coupling medium, or the air (airborne ultrasound). Limits for exposure from each mode should be treated separately.

9.2.2.1 Airborne ultrasound

A number of human-exposure limits for airborne acoustic radiation have been proposed and these are summarized in Tables 18 and 19. From the results of her studies Grigor'eva concluded: "The experiments lead one to believe that airborne ultrasound is considerably less hazardous to man in comparison with audible sound. Also bearing in mind the data available in the literature, 120 dB may be adopted as an acceptable limit for the acoustic pressure for airborne ultrasound. The possibility of raising this level should be tested experimentally." (Grigor'eva, 1966a, b). In her work on both audible and inaudible components of airborne ultrasound, Grigor'eva did not propose any exposure-time limits for her suggested values for acceptable limits of acoustic pressure.

Acton (1968) proposed a criterion below which auditory damage and/or subjective effects were unlikely to occur as a result of human exposure to airborne noise from industrial ultrasonic sources over a working day. He based his criterion on the belief that it is the high audible frequencies present in the noise from ultrasonic machines, and not the ultrasonic frequencies themselves, that are responsible for producing subjective effects. He extended this criterion to produce a tentative estimate for an extension to damage risk criteria, giving levels of 110 dB in the one-third octave bands centred on 20, 25, and 31.5 kHz. This extended criterion was chosen to cover the possible occurrence of: (a) generation of first-order subharmonics of potentially hazardous levels in the audible frequency range, and (b) subjective effects arising from subharmonic distortion products occurring at and below 16 kHz. Acton (1974) reported that additional data obtained for industrial exposures confirmed that the levels set in the proposed criterion were at approximately the right level, and that there did not seem to be any necessity to amend them.

Parrack (personal communication, 1969) proposed a criterion for a standard having acceptable levels of high-frequency airborne sound low enough to: (a) prevent adverse bioeffects (subjective effects), and (b) protect the

- 137 -

Table 18. Exposure limits (dB) for airborne acoustic energy
at the workplace[a]

Sound pressure level within one-third octave band
(dB relative to 20 µPa)

Mid-frequency of one-third octave band (kHz)	Jpn. Min. Lab. (1971)	Acton (1975)	USSR St. (1975)	USAF (1976)	Dept H & W Canada (1980b)	Sweden (1978)	ACGIH (1981)	IRPA draft (1981)
8	90	75			80		80	80
10	90	75			80		80	80
12.5	90	75	75	85	80		80	80
16	90	75	85	85	80		80	80
20	110	75	110	85	80	105	105	80
25	110	110	110	85	110	110	110	110
31.5	110	110	110	85	110	115	115	110
40	110	110	110	85	110	115	115	110
50	110		110		110	115	115	110

[a] For total ultrasound exposure exceeding 4 h/day.

Table 19. Permitted increase in sound pressure levels (SPLs) in Table 18
at workplaces in the vicinity of ultrasound sources

	Total ultrasound exposure time (per day)	Permitted rise in SPL	Total ultrasound exposure time (per day)	Permitted rise in SPL
USSR St. (1975)	1 - 4 h	+6	5 - 15 min	+18
	1/4 - 1 h	+12	1 - 5 min	+24
Sweden (1978)	1 - 4 h	+3		
	0 - 1 h	+9		
IRPA (draft) (1981)	1 - 4 h	+3		
	0 - 1 h	+9		

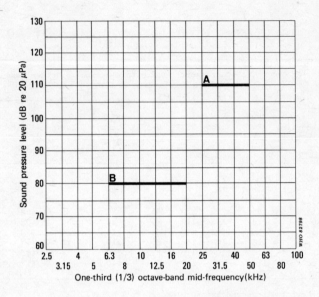

Fig. 9. Recommended exposure limits for airborne ultrasound (From: Canada, Department of National Health and Welfare, 1980b).

A = line representing the maximum sound pressure level for frequencies 25 kHz and above.
B = line representing the maximum sound pressure level for frequencies 20 kHz and below.

NOTE: The nominal centre frequency of 20 kHz has a one-third octave passband from 17.8-22.4 kHz, and the nominal centre frequency of 25 kHz has a one-third octave passband of 22.4-28.2 kHz.

hearing of persons exposed to noise from ultrasonic equipment and machines over a working period of 8 h per day (nominally) for 5 or 5 1/2 days each week. The criterion was based on Parrack's experimental findings of temporary threshold shifts in hearing levels at subharmonic frequencies for several subjects exposed to high frequency sound. The American Conference of Governmental Industrial Hygienists used Parrack's criterion for their ultrasound exposure levels (ACGIH, 1981).

Ultrasound noise is limited to 85 dB per one-third octave by the US Air Force (US Air Force, 1976) for frequencies in the range of 12.5-40 kHz.

The USSR has maximum sound pressure levels to limit
exposure of workers in the vicinity of ultrasound sources
(USSR State Committee for Standards, 1975). The levels are
divided into three frequency ranges by one-third octave
bands. The maximum sound pressure level for the corresponding
geometric frequency mean by one-third octave band is 75 dB for
12.5 kHz, 85 dB for 16 kHz, and 110 dB for 20 kHz (ILO, 1977).
The levels stated therein may be increased, when the total
duration of exposure does not exceed 4 h per day, in
accordance with Table 19.

The National Board of Occupational Safety and Health in
Sweden (Sweden, 1978) has issued directions concerning
airborne ultrasound exposure in the frequency range of 20-200
kHz. The levels are also divided into 3 frequency ranges by
the mid-frequency of the one-third octave band of 20, 25, and
>31 kHz. The maximum sound pressure levels are given in
Table 18 for exposure durations exceeding 4 h per day and in
Table 19 for exposure times of less than 4 h.

The Department of National Health and Welfare, Canada,
(1980b) requires that the one-third octave band levels (lines
A and B of Fig. 9) be used as the exposure limits for airborne
ultrasound, because adverse health effects seem to arise from
"single frequency" components. One-third octave band filters
appear to be narrow enough in frequency band width for the
required analysis. These filters are readily available and
can be obtained with flat response networks up to higher
frequencies. The 6.3 kHz, one-third octave band has been
chosen to begin specifying criteria levels, because no adverse
(subjective) effects have been found below this frequency.

In Japan, noise levels from ultrasonic welders have been
regulated at values of less than 90 dB for frequencies of less
than 16 kHz (one-third octave band) and less than 110 dB for
frequencies higher than 20 kHz (one-third octave band), under
the provision of a circular of the Japanese Ministry of Labour
(Japanese Ministry of Labour, 1971). There are many Japanese
automobile factories in which more than 100 ultrasonic welders
are in operation.

The International Radiation Protection Association (IRPA,
1981) has drafted the first international limits for human
exposure to airborne acoustic energy having one-third octave
bands with mid frequencies from 8 to 50 kHz. Tables 18 and 19
indicate the proposed IRPA limits for occupational exposure.
This proposal is similar to the standards existing in a number
of countries. The document incorporating the proposal also
contains a scientifically based rationale for the limits. The
IRPA (IRPA, 1981) has also proposed a set of exposure limits
for exposure of the general population to airborne acoustic
energy. Table 20 gives the details of this proposal.

Table 20. Limits of continuous exposure of the general
population to airborne acoustic energy[a]

Mid-frequency of one-third octave band (kHz)	SPL within one-third octave (dB re: 20 µPa)	
	Day	Night
8	41	31
10	42	32
12.5	44	34
16	46	36
20	49	39
25	110	110
31.5	110	110
40	110	110
50	110	110

[a] From: IRPA (1981).

9.3 Specific Protective Measures

9.3.1 Diagnostic ultrasound

Reviews of current knowledge on biological effects and applications of diagnostic ultrasound (section 5.3.1) suggests that:

(a) Ultrasonic output information should be supplied to the user. This information should include total power, SATA intensity, SPTA intensity, SPTP intensity, SPPA intensity, pulse length, and pulse repetition frequency, as applicable. Criteria for imaging effectiveness should also be developed and disseminated. Such criteria would help the user evaluate benefit versus risks and aid the user in keeping the output of ultrasonic equipment as low as practicable, consistent with obtaining the necessary diagnostic information.

Some procedures for making intensity measurements have been specified (AIUM-NEMA, 1981). Manufacturers and users should strive to develop meaningful standardized techniques to evaluate imaging effectiveness.

(b) Output levels approaching the lower limits of those used in therapy should not be employed for diagnostic purposes, unless they can be justified on the basis of obtaining necessary information not otherwise obtainable. Equipment with output levels exceeding the lower limits of those used in therapy (i.e., SATA intensities above 100 mW/cm²) should include instruments for monitoring both exposure level and exposure time as recommended in the Canadian safe-use guidelines (Canada, Department of National Health and Welfare, 1980a).

(c) More information is needed with regard to effects of exposure from pulsed units before guidelines concerning SPPA or SPTP intensities can be developed. There is evidence that diagnostic pulse-echo ultrasound causes biological damage to certain tissues. This effect apparently is a result of some form of cavitation activity and occurs because of microscopic gas-filled spaces within these tissues. The damage is closely correlated with the temporal peak intensity rather than the time-averaged value (Carstensen, 1982).

(d) In general, equipment should be designed with adjustable controls so that the operator can use the minimum acoustic exposure required to image or obtain other information concerning the organ of interest in each patient. These adjustable controls are especially needed for fetal Doppler equipment because: (i) fetal monitoring can involve extremely long exposure times (of the order of hours or days when a stationary transducer is strapped to the mother's abdomen); (ii) this application involves direct exposure of the fetus. It should be noted that it is technically and commercially feasible to build effective fetal Doppler equipment with output levels below SATA intensities of 10 mW/cm² (JIS, 1979).

(e) Diagnostic ultrasound should be used for human exposure only when there is a valid medical reason. Individuals, especially when pregnant, should not be exposed for commercial demonstration or for routine imaging to produce test images when equipment is being serviced (AAPM, 1975).

(f) Quality control and testing programmes to ensure equipment performance specifications are met should be adopted by manufacturers and users. Quality control procedures for maintaining diagnostic ultrasound at a high level of efficiency have been described by Goldstein (1982).

9.3.2 Therapeutic ultrasound

The reviews of biological effects (section 6), applications (section 5.3.2), and instrumentation (section 4) related to therapeutic ultrasound suggest that:

(a) Accuracy specifications for the acoustic output power and the timer are needed, because both directly affect the dose delivered to the patient.

(b) There are arguments for and against setting upper limits to the intensity of the beam of an ultrasound therapy device. It should be remembered that physiotherapists want to produce an effect on the region of injury, and they require an appropriate amount of ultrasound energy to achieve this aim. An upper limit might be construed as a "safe level" for exposure, thus encouraging its use. Above 3 W/cm², the heat generated is generally unbearable for most patients; moreover such an intensity has been reported to retard bone growth (Kolar et al., 1965). In addition, cavitation, which may cause significant tissue damage, is increasingly possible at intensities above this level.

(c) (i) Because fetal abnormalities and reduced suckling weight have been observed after pregnant mice have been exposed at therapeutic intensities, no pregnant patient should receive ultrasound therapy in a way that is likely to expose the fetus directly or indirectly. At present, it is common to give ultrasound therapy to pregnant patients for lower back pain. This practice should definitely be discouraged. (ii) It is not advisable to use ultrasound over the vertebral column, especially following laminectomies, or when any anaesthetized areas are involved. (iii) Care should be taken when epiphyseal lines in children are exposed to ultrasound, especially when these regions are still at the growing stage. (iv) Care should be exercised, when treating peripheral vascular disease within extremities, because with diminished sensation and lack of blood circulation, the patient may not detect overexposure to ultrasound.

(d) Ultrasound exposure close to a strong reflecting surface such as bone may lead to the formation of standing waves, with the possibility of producing blood-flow stasis and related effects. Endothelial damage to the blood vessels may ensue, if such stasis occurs for extended periods of time. In therapy, the ultrasound transducer should be moved over the region of injury to minimize harmful effects from standing waves and possible cavitation.

(e) Operators of therapeutic ultrasound devices should avoid exposure in two main areas: (i) large blood pools (e.g., heart, spleen); (ii) reproductive organs (e.g., testes, ovaries, pregnant uterus).

Most of the precautions listed above are not absolute and refer to the direct exposure of the site mentioned. They are on the conservative side and may change as more data become available. While there would be a contraindication for therapy with high SATA intensities in a case of peripheral vascular insufficiency in the leg, this would not mean that the same patient could not be treated with ultrasound for a "frozen" shoulder. Likewise, though the pregnant uterus should not be

directly exposed to therapeutic ultrasound, applications to
other parts of the body, such as an extremity, should not
result in any significant exposure of the fetus.

(f) Patient exposure can and should be minimized by: (i)
testing patient skin sensation prior to application of
ultrasound (if patients have sensory paralysis and are unable
to differentiate between hot and cold, an alternative type of
treatment should be given, since they would not be able to
detect overexposure; the same criterion applies to treating
patients when anaesthetized areas are involved); (ii) using
the minimum effective exposure (i.e., ultrasound power and
duration of exposure); (iii) keeping the energized transducer
moving slowly over the treatment region to minimize the risk
of "hot spots" (undue temperature elevation in tissue
receiving excessive exposure); (iv) reducing the ultrasound
power level, if a mild tingling sensation or pain is felt in
the treatment region (such a sensation may be an indication
that there is overheating within the treatment region, and
significant damage to the tissue could occur if this sensation
is allowed to continue); (v) ensuring that the operator is
present to terminate the treatment if the patient shows the
least sign of distress; (vi) calibrating equipment used for
treatment purposes to provide the operator with the capability
of delivering acoustic intensities that are below levels at
which adverse biological or subjective effects have been
reported.

(g) Well-designed controlled clinical trials should be
carried out to evaluate the effectiveness of ultrasound
treatments. By this means, ineffective treatments may be
identified and either eliminated or modified so that they
become efficacious.

(h) Operator exposure can be minimized by: (i) not
touching the face of the transducer or applicator when it is
emitting ultrasound; and (ii) not immersing any part of the
operator's body in the water-bath while ultrasound is being
generated.

9.3.3 Industrial, liquid-borne, and airborne ultrasound

The reviews of industrial, liquid-borne, and airborne
ultrasound sources (section 5.1, 5.2, 7) and effects suggest
that:

(a) Exposure levels should be minimized and certainly be
below levels at which adverse biological or subjective effects
have been reported.

(b) Persons exposed to high levels of noise associated
with ultrasonic equipment should be protected either by
wearing devices such as earmuffs, or by acoustic barriers
constructed around the equipment to reduce the noise levels.

(c) Direct contact exposure to high intensities of liquid-borne ultrasound should be avoided. For example, operators should not place their hands in ultrasonic cleaning tanks during operation. Warning signs to this effect should be placed at suitable locations.

(d) In burglar alarm systems, the ultrasonic source itself should be switched off, instead of only the alarm, when the system is not in use.

(e) Care should be taken that ultrasonic transmitters used for smoke coagulation are located so that they do not expose workers nearby.

9.3.4 General population exposure

The general population may be exposed to ultrasound from a number of sources. Some of these might be grouped as:

(a) Consumer sources, exemplified by ultrasonic cleaners, remote control devices, sonar devices, dog control and repelling devices, distance-measuring devices for cameras, etc.

(b) Public sources, exemplified by sources in public areas such as door openers, burglar alarms, devices for bird and rodent control, etc.

Of the devices mentioned above, only the ultrasonic cleaners, dog repelling devices, and burglar alarms are likely to cause any concern. Consumer sources are often handled by a limited number of persons, who should obtain pertinent information concerning function, use, and possible risks. Manufacturers should only market devices in which the operational intensities are considered safe to use and comply with standards current at the time of manufacture (section 9.2.2.1). Unnecessary use should be avoided.

In addition to these protective measures, ultrasound sources used near the general population should be properly labelled with appropriate protective information; the radiation area should be marked so that people will avoid staying in radiated areas for prolonged periods.

9.4 Education and Training

An educational programme on the safe use of ultrasound is one of the most important aspects of protection. Such a programme entails education of the general population and training of users of ultrasound devices. The development of educational materials should be a key aspect of such a programme.

A document outlining safe-use guidelines for device operators should include the following:
   (a) care and use of ultrasound equipment;
   (b) measurement and calibration of the equipment;
   (c) operator training programme;
   (d) a summary of biological effects that may arise from ultrasound exposure;
   (e) information on how patient doses can be reduced by lowering exposure where practical;
   (f) contraindications - when not to use ultrasound;
   (g) recommended exposure limits;
   (h) safe operating procedures.

Publications containing such information are available (AAPM, 1979; Canada, Department of National Health and Welfare, 1980a, b; Goldstein, 1982).

Many applications of ultrasound involve control of complicated equipment. In diagnostic imaging procedures, for example, the skill of the operator has a great influence on the diagnostic efficiency on the time required to make the examination. The operator has to select scanning planes and instrument parameters in an interactive process dependent on the actual findings. Incorrect control of the ultrasound scanner can result in two different forms of risk:
   (a) excessive exposure of the patient to ultrasound radiation because of long exposure times;
   (b) incorrect diagnosis, which in turn might lead to repeated exposures.

The obvious solution is well-planned and supervised education and training of all personnel working with ultrasound radiation.

REFERENCES

AAPM (1975) Statement on the use of diagnostic ultrasound instrumentation on humans for training, demonstration and research, General Medical Physics Committee of the AAPM. Med. Phys., 2(1): 38.

AAPM (1979) Ultrasound instrument quality control procedures. Maryland, American Association of Physicists in Medicine, Cleaveland, Ohio, Chemical Rubber Publishing Co., p. 45 (CRP Report Series - Report 3).

AAPM (1980) Pulse echo ultrasound imaging systems: Performance tests and criteria, New York, American Institute of Physics (American Association of Physicists in Medicine Report No. 8).

ABDULLA U., CAMPBELL, S., DEWHURST, C.J., TALBERT, D., LUCAS, M., & MULLARKEY, M. (1971) Effect of diagnostic ultrasound on maternal and fetal chromosomes. Lancet, 2: 829-831.

ACGIH (1981) Threshold limit values for physical agents. Cincinnati, Ohio, American Conference of Governmental Industrial Hygienists, USA.

ACTON, W.I. (1968) A criterion for the prediction of auditory and subjective effects due to airborne noise from ultrasonic sources. Ann. occup. Hyg., II: 227-234.

ACTON, W.I. (1973) The effects of airborne ultrasound and near ultrasound. In: International Congress on Noise as a Public Health Problem. Dubrovnik, 14-18 May 1973, pp. 349-359.

ACTON, W.I. (1974) The effects of industrial airborne ultrasound on humans. Ultrasonics, 12: 124-128.

ACTON, W.I. (1975) Exposure criteria for industrial ultrasound. Ann. occup. Hyg., 18: 267-268.

ACTON, W.I. & CARSON, M.B. (1967) Auditory and subjective effects of airborne noise from industrial ultrasonic sources. Br. J. ind. Med., 24: 297-304.

ACTON, W.I. & HILL, C.E. (1977) Hazards of industrial ultrasound. Protection, 14(19): 12-17.

AHRLIN, U. & OHRSTROM, B. (1978) Medical effects of environmental noise on humans. J. Sound Vib., 59: 79-87.

AIUM (1974) 100 millimeter test object including standard procedure for its use. Washington, DC, American Institute of Ultrasound in Medicine.

AIUM (1978a) American Institute of Ultrasound in Medicine bioeffects statement. Reflections, 4(4): 311 (also see "Who's afraid of a hundred milliwatts per square centimeter (100 mW/cm2, SPTA)?", brochure prepared by American Institute of Ultrasound in Medicine Bioeffects Committee, Washington, DC).

AIUM (1978b) American Institute of Ultrasound in Medicine standard on presentation and labeling of ultrasound images. Reflections, 4: 70-75.

AIUM (1979) Standard specification of echoscope sensitivity and noise level including recommended practice for such measurements, Washington, DC, American Institute of Ultrasound in Medicine.

AIUM (1980) Recommended nomenclature: Physics and engineering, Washington, DC, American Institute of Ultrasound in Medicine.

AIUM (1981) American Institute of Ultrasound in Medicine standard for transducer characterization, Washington, DC, American Institute of Ultrasound in Medicine.

AIUM-NEMA (1981) AIUM-NEMA safety standard for diagnostic ultrasound equipment (Draft V, January 27, 1981), Washington, DC, American Institute of Ultrasound in Medicine.

AKAMATSU, N. (1981) Ultrasound irradiation effects on pre-implantation embryos. Acta Obstet. Gynaecol. Jpn., 33(7): 969-978.

AKAMATSU, N. & SEKIBA, K. (1977) Symposium on recent studies in the safety of diagnostic ultrasound. Ultrasound irradiation effect on pre-implanted embryos. Jpn. J. Med. Ultrason., 4: 274-278.

AKAMATSU, N., NIWA, K., SEKIBA, K., & UTSUMI, K. (1977) Ultrasound irradiation effect on embryos (6). Effect of continuous ultrasound irradiation on pre-implanted rat embryo (2). Nippon Choompa Igakkai Koen-Rombunshu, 32: 151-152.

AKOPYAN, V.B. & SARVAZYAN, A.P. (1979) Investigations of mechanisms of action of ultrasound on biological media and object. Sov. Phys. Acoust., 25(3): 262-263.

AL-HASHIMI, A.H.M. & CHAPMAN, I.V. (1981) Modification of ultrasound-induced changes in mammalian cells by increased viscosity of medium and increased ambient pressure. Int. J. Radiat. Biol., 38: 11-19.

ALLEN, C.H., BRINGS, H., & RUDNICK, I. (1948) Some biological effects of intense high frequency airborne sound. J. Acoust. Soc. Am., 20(1): 62-65.

AMIN, A., FOSTER, K.R., TERNES, J., & TAKASHIMA, S. (1981) Lack of effect of pulsed ultrasound on the mammalian EEG. Aviat. Space Environ. Med., 52: 604-607.

ANDERSON, D.W. & BARRETT, J.T. (1979) Ultrasound: A new immunosuppressant. Clin. Immunol. Immunopathol., 14: 18-29.

ANDERSON, D.W. & BARRETT, J.T. (1981) Depression of phagocytosis by ultrasound. Ultrasound Med. Biol., 7: 267-273.

ANDERSON, G.H., HELLUMS, J.D., MOAKE, J.L., & ALFREY, C.P., Jr (1978) Platelet response to shear stress: Changes in serotonin uptake, serotonin release, and ADP induced aggregation. Thromb. Res., 13: 1039-1047.

ANGELUSCHEFF, Z.D. (1954) Ultrasonics and progressive deafness. J. Acoust. Soc. Am., 26: 942 (abstract).

ANGELUSCHEFF, Z.D. (1955) Sonochemistry and the organ of hearing. J. Acoust. Soc. Am., 27: 1009.

ANGELUSCHEFF, Z.D. (1967) Ultrasonics, resonance and deafness. Rev. Laryngol. Otol. Rhinol., July-August.

APFEL, R.E. (1981) Acoustic cavitation. In: Edmonds, P.D., ed. Methods of experimental physics - ultrasonics, New York, London, Toronto, Sydney, and San Francisco, Academic Press, Vol 19, pp. 356-411.

ARSLAN, M., GREPALDI, G., GRANDESSO, R., MOLINARI, G.A., MUGGEO, M., & RICCI, V. (1973) Direct ultrasonic irradiation of the hypophysis, Padua, Piccin Medical Books.

AŠBEL, Z.Z. (1965) [The effect of ultrasound and high frequency noise upon the blood sugar level.] Gig. Tr. Prof. Zabol., 9: 29-33 (in Russian) (Occup. Health Safety Abstr.).

ASSOCIATION FRANCAISE DE NORMALISATION (1963) Appareils à ultrasons, Paris (Norme française NF C 74-306).

ASSOCIATION FRANCAISE DE NORMALISATION (1982) Appareils à ultrasons utilisés en diagnostic, Paris (Norme française NF C 74-335).

BALAMUTH, L. (1967) The application of ultrasonic energy in the dental field. In: Brown, B. & Gordon, D., ed. Ultrasonic techniques in biology and medicine, London, Iliffe, pp. 194-205.

BAMBER, J.C., HILL, C.R., & KING, J.A. (1981) Acoustic properties of normal and cancerous human liver - II. Dependence of tissue structure. Ultrasound Med. Biol., 7: 135-144.

BANG, J. (1972) The intensity of ultrasound in the uterus during examination for diagnostic purposes. Acta pathol. microbiol. Scand., Section A., 80: 341-344.

BANG, J. & NORTHEVED, A. (1970) Ultrasonic equipment for application of ultrasound with high effect to animals used for experiments. Acta pathol. microbiol. Scand., Section A., 78: 219-230.

BANG, J. & NORTHEVED, A. (1972) A new ultrasonic method for transabdominal amniocentesis. Am. J. Obstet. Gynec., 114: 599-601.

BARNETT, S.B. (1979) Bioeffects of pulsed ultrasound. Austral. Phys. Sci. Med., 2-7: 397-403.

BARNETT, S.B. & KOSSOFF, G. (1977) Negative effect of long duration pulsed irradiation on the mitotic activity in regenerating rat liver. In: White, D. & Brown, R.E., ed. Ultrasound in Medicine, New York and London, Plenum Press, Vol 3B, pp. 2033-2044.

BARTH, G. & WACHSMANN, F. (1949) Biological effects of ultrasonic therapy. In: Report of the Erlangen Ultrasound Congress, Stuttgart, pp. 162-205.

BEKHOME, A.K. (1977) [Leukergia reaction in patients suffering from chronic tonsillitis before and after ultrasound therapy.] Vestn. Otorinolaringol., 4: 62-65 (in Russian with English abstract).

BELEWA-STAIKOWA, R. & KRASCHKOWA, A.M. (1967) Effects of biophysical factors on the redox processes and biological oxidation. Effect of ultrasonics on the protein content and transaminase activity of organs. Radiobiol. Radiother., 8: 655-662.

BENDER, L.F., JANES, J.M., & HERRICK, J.F. (1954) Histologic studies following exposure of bone to ultrasound. Arch. Phys. Med. Rehab., 35: 555-559.

BERNAT, R., HRYNIEWIECKI, L., & STRABURZYNSKI, G. (1966a) [Influence of ultrasonics on the behaviour of proteins and simpler nitrogen components in the blood and certain organs, and on blood osmolarity.] Acta Physiol. Pol., 17(2): 193-203 (in Polish).

BERNAT, R., HRYNIEWIECKI, L., & STRABURZYNSKI, G. (1966b) [Proteins and amino acids in the cataract induced by ultrasonics.] Acta Physiol. Pol., 17(2): 225-235 (in Polish).

BERNSTEIN, R.L. (1969) Safety studies with ultrasonic Doppler technique - a clinical follow-up of patients and tissue culture study. Obstet. Gynecol., 34(5): 707-709.

BEYER, R.T. & LETCHER, R. (1969) Physical ultrasonics, New York, Academic Press.

BINDAL V.N. & KUMAR A. (1979) Effect of the density of liquids used on the sensitivity of float method for ultrasonic power measurement. J. pure. app. Ultrason., 1: 69-71.

BINDAL V.N. & KUMAR, A. (1980) Measurement of ultrasonic power with a fixed path radiation pressure float method. Acustica, 46: 223-225.

BINDAL, V.N., SINGH, V.R., & SINGH, G. (1980) Acoustic power measurement of medical ultrasonic probes using a strain gauge technique. Ultrasonics, 18: 28-31.

BORRELLI, M.J., BAILEY, K.I., & DUNN, F. (1981) Early ultrasonic effects on mammalian CNS structure (chemical synapses). J. Acoust. Soc. Am., 69(5): 1514-1517.

BOUCHER, R.M. & KRUETER, J. (1968) The fundamentals of the ultrasonic atomization of medicated solutions. Ann. Allerg., 26: 591-600.

BOYD, E., ABDULLA, U., DONALD, I., FLEMING, J.E.E., HALL, A.J., & FERGUSON-SMITH, M.A. (1971) Chromosomal breakage and ultrasound. Br. med. J., 2: 501-502.

BRENDEL, K., MOLKENSTRUCK, W., & REIBOLD, R. (1978) Targets for ultrasonic power measurements. In: 3rd European Congress on Ultrasonics in Medicine, Bologna, Oct. 1-5, 1978, pp. 473-476.

BROWN, G.G. (1967) Airborne ultrasonics - their effects upon humans. Can. Hosp., 44: 55-56.

BROWN, C.H., LEVERETT, L.B., LEWIS, C.W., ALFREY, C.P., & HELLUMS, J.D. (1975) Morphological, biochemical and functional changes in human platelets subjected to shear stress. J. lab. clin. Med., 86: 462-471.

BROWN, N., GALLOWAY, W.D., MONAHAN, J.C., & FISHER, B. (1979) Postnatal behavior and development. In utero exposure of mice to ultrasound and microwave radiation. In: Proceedings of the Fifth FDA Science Symposium on "Methods for Predicting Toxicity", Arlington, VA, Oct. 10-12, 1979.

BROWN, N., GALLOWAY, W.D., & HENTON, W.W. (1981) Reflex development following in utero exposure to ultrasound. In: Proceedings of the American Institute of Ultrasound in Medicine, San Francisco, CA, Aug. 17-21.

BUCKTON, K.E. & BAKER, N.V. (1972) An investigation into possible chromosome damaging effects of ultrasound on human blood cells. Br. J. Radiol., 45: 340-342.

BUNDY, M.L., LERNER, J., MESSIER, D.L., & ROONEY, J.A. (1978) Effects of ultrasound on transport in avian erythrocytes. Ultrasound Med. Biol., 4: 259-262.

BURR, J.G., WALD, N., PAN, S., & PRESTON, K. Jr (1978) The synergistic effect of ultrasound and ionizing radiation on human lymphocytes. In: Evans, H.J. & Lloyd, D.C., ed. Mutagen-induced chromosome damage in man, New Haven, Connecticut, Yale University Press, pp.120-128.

BYALKO, N. (1964) Certain biochemical abnormalities in workers exposed to high frequency noise. Excerpta Med., 17: 570.

CACHON, J., CACHON, M., & BRUNETON, J.H. (1981) An ultrastructural study of the effect of very high frequency ultrasound on a microtubular system. Biol. Cell, 40: 69.

CANADA, DEPARTMENT OF NATIONAL HEALTH AND WELFARE (1980a) Guidelines for the safe-use of ultrasound, Part I - Medical and paramedical applications. Safety Code-23 (Health and Welfare, Canada, Publication, 80-EHD-59).

CANADA, DEPARTMENT OF NATIONAL HEALTH AND WELFARE (1980b) Guidelines for the safe use of ultrasound, Part II - Industrial and commercial applications. Safety Code-24. (Health and Welfare, Canada, Publication, 80-EHD-60).

CANADA, DEPARTMENT OF NATIONAL HEALTH AND WELFARE (1981) Ultrasound therapy devices regulation. Canada Gaz., Part II, 115(8): 1121-1126.

CARSON, P.L. (1980) Diagnostic ultrasound emissions and their measurement. In: Fullerton, G. & Zagzebski, J., ed. 1980 AAPM Summer School Proceedings, New York, American Institute of Physics.

CARSON, P.L., FISCHELLA, P.R., & OUGHTON, T.V. (1978) Ultrasonic power and intensities produced by diagnostic ultrasound equipment. Ultrasound Med. Biol., 3(4): 341-350.

CARSON, T.E. & FISHMAN, S. (1976) Biological effects of ultrasound: skin and cutaneous blood vessels. Proc. West. Pharmacol. Soc., 19: 36-39.

CARSTENSEN, E.L. (1982) Biological effects of low-temporal-average intensity pulsed ultrasound. Bioelectromagnetics, 3: 147-156.

CARSTENSEN, E.L., BECROFT, S.A., LAW, W.L., & BARBEE, D.B. (1981) Finite amplitude effects on the thresholds for lesion production in tissues by unfocused ultrasound. J. Acoust. Soc. Am., 70(2): 302-309.

CATALDO, F.L., MILLER, M.W., & GREGORY, W.D. (1973) A description of ultrasonically-induced chromosomal anomalies in Vicia faba. Radiat. Bot., 13: 211-213.

CHAN, A.K., SIGELMAN, R.A., & GUY, A.W. (1974) Calculations of therapeutic heat generated by ultrasound in fat-muscle-bone layers. IEEE Trans. Biomed. Eng., 21: 280-284.

CHAPMAN, I.V. (1974) The effect of ultrasound on the potassium content of rat thymocytes in vitro. Br. J. Radiol., 47: 411-415.

CHAPMAN, I.V., MACNALLY, N.A., & TUCKER, S. (1980) Ultrasound-induced changes in rates of influx and efflux of potassium ions in rat thymocytes in vitro. Ultrasound Med. Biol., 6: 47-58.

CHATER, B.V. & WILLIAMS, A.R. (1977) Platelet aggregation induced in vitro by therapeutic ultrasound. Thromb. Haemos., 38: 640-651.

CHAUSSY, C., BRENDEL, W., & SCHNIEDT, E. (1980) Extra-corporeally induced destruction of kidney stones by shock waves. Lancet, 2: 1265-1268.

CHILD, S.Z., CARSTENSEN; E.L., & SMACHLO, K. (1980) Effects of ultrasound on Drosophila - I. Killing of eggs exposed to travelling and standing wave fields. Ultrasound Med. Biol., 6: 127-130.

CHILD, S.Z., CARSTENSEN, E.L., & LAM, S.K. (1981a) Effects of ultrasound on Drosophila - III. Exposure of larvae to low-temporal-average-intensity pulsed irradiation. Ultrasound Med. Biol., 7: 167-173.

CHILD, S.Z., CARSTENSEN, E.L., & DAVIS, H.T. (1981b) Tests for "miniature flies" following exposure of Drosophila melanogester larvae to diagnostic levels of ultrasound. Exp. cell Biol., 48: 461-466.

CHILD, S.Z., HARE, J.D., CARSTENSEN, E.L., VIVES, B., DAVIS, J., ALDER, A., & DAVIS, H.T. (1981c) Test for the effects of diagnostic levels of ultrasound on the immune response of mice. Clin. Immunol. Immunopathol., 18: 299-302.

CHORAZAK, T. & KONECKI, J. (1966) The effect of ultrasonics on the content of protein bound SH and SS groups in the mouse epidermis. Acta Histochem., 25: 127-132.

CIARAVINO, V., FLYNN, H.G., & MILLER, M.W. (1981) Pulsed enhancement of acoustic cavitation: a postulated model. Ultrasound Med. Biol., 7: 159-166.

CLARKE, P.R. & HILL, C.R. (1969) Biological action of ultrasound in relation to the cell cycle. Exp. Cell Res., 58: 443-444.

CLARKE, P.R. & HILL, C.R. (1970) Physical and chemical aspects of ultrasonic disruption of cells. J. Acoust. Soc. Am., 50: 649-653.

CLARKE, P.R., HILL, C.R., & ADAMS, K. (1970) Synergism between ultrasound and X-rays in tumour therapy. Br. J. Radiol., 43: 97-99.

COAKLEY, W.T. (1978) Biophysical effects of ultrasound at therapeutic intensities. Physiotherapy, 64(6): 166-169.

COAKLEY, W.T. & NYBORG, W.L. (1978) Cavitation; dynamics of gas bubbles; applications. In: Fry, F.J., ed. Ultrasound: Its application in medicine and biology, Amsterdam, Elsevier Publishing Co., pp. 77-159.

COAKLEY, W.T., HAMPTON, D., & DUNN, F. (1971) Quantitative relationships between ultrasonic cavitation and effects upon amoebae at 1 MHz. J. Acoust. Soc. Am., 50: 1546-1553.

COLES, R.R.A. & KNIGHT, J.J. (1965) The problems of noise in the Royal Navy and Royal Marines. J. Laryngol. Otol., 79: 131-147.

COMBES, R.D. (1975) Absence of mutation following ultrasonic treatment of Baccillus subtilis cells and transforming deoxyribonucleic acid. Br. J. Radiol., 48: 306-311.

CONNOLLY, W. & FOX, F.E. (1954) Acoustic cavitation thresholds in water. J. Acoust. Soc. Am., 26: 843-848.

CRABTREE, R.B. & FORSHAW, E.E. (1977) Exposure to ultrasonic cleaner noise in the Canadian Forces, Ottawa (Dept of National Defence, DCIEM Technical Report No. 77 X 45).

CRUM, L.A. & HANSEN, G.M. (1982) Growth of air bubbles in tissue by rectified diffusion. Phys. Med. Biol., 27: 413-417.

CURTO, K.A. (1975) Early postpartum mortality following ultrasound radiation. In: Annual Conference, American Institute of Ultrasound in Medicine, Winston-Salem, 4-9 Oct., pp. 535-536 (abstract).

DANIELS, S., PATON, W., & SMITH, E.B. (1979) Ultrasonic imaging system for the study of decompression induced air bubbles. Undersea biomed. Res., 6: 197-209.

DANNER, P.A., ACKERMAN, E., & FRINGS, H.W. (1954) Heating of haired and hairless mice in high intensity sound fields from 6 to 22 kc. J. Acoust. Soc. Am., 26: 731.

DAVID, H., WEAVER, J.B., & PEARSON, J.F. (1975) Doppler ultrasound and fetal activity. Br. med. J., 2: 62-64.

DAVIES, H., SCHWARTZ, R., PFISTER, R., & BARNES, F. (1974) Transmitted ultrasound for relief of obstruction in ureters and arteries: current status. J. clin. Ultrasound, 2: 217-220.

DAVIES, H., BEAN, W.J., & BARNES, F.S. (1977) Breaking up of residual gallstones with an ultrasonic drill. Lancet, 2(8032): 278-279.

DAVIS, H. (1948) Biological and physiological effects of ultrasonics. J. Acoust. Soc. Am., 20: 605-607.

DAVIS, H., ed. (1958) Auditory and nonauditory effects of high intensity noise, Pensacola, Florida, Central Institute for the Deaf and Naval School of Aviation.

DeREGGI, A.S., ROTH, S.C., KENNEY, J.M., EDELMAN, S., & HARRIS, G. (1978) Polymeric ultrasonic probe. J. Acoust. Soc. Am., 64(1): 55 (abstract).

DeREGGI, A.S., ROTH, S.C., KENNEY, J.M., EDELMAN, S., & HARRIS, G. (1981) Piezoelectric polymer probe for ultrasonic applications. J. Acoust. Soc. Am., 69(3): 853-859.

DESCH, E.H., SPROULE, D.A., & DAWSON, W.J. (1946) The detection of cracks in steel by means of supersonic waves. J. Steel Inst., 153: 319.

DEWITZ, T.S., MARTIN, R.R., SOLIS, R.T., HELLUMS, J.D., & McINTIRE, L.V. (1978) Microaggregate formation in whole blood exposed to shear stress. Microvasc. Res., 16: 263-271.

DEWITZ, T.S., McINTIRE, L.V., MARTIN, R.R., & SYBERS, H.D. (1979) Enzyme release and morphological changes in leukocytes induced by mechanical trauma. Blood Cells, 5: 499.

DICKSON, J.A. & ELLIS, H.A. (1974) Stimulation of tumor cell dissemination by raised temperature (42 °C) in rats with transportable Yoshida tumors. Nature (Lond.), 248: 354-358.

DOBROSERDOV, V.K. (1967) [The effects of low frequency ultrasonic and high frequency sound waves on workers.] Gig. i Sanit., 32: 17-21 (in Russian).

DUMONTIER, A., BURDICK, A., EWIGMAN, B., & FAHIM, M.S. (1977) Effects of sonication on mature rat testes. Fertil. Steril., 28(2): 195-204.

DUNN, F. (1965) Ultrasonic absorption by biological materials. In: Kelly, E., ed. Ultrasonic energy. Illinois, University of Illinois Press, pp. 51-65.

DUNN, F. & COAKLEY, W.T. (1972) Interaction of ultrasound and microorganisms in suspension. In: Reid, J.M. & Sikov, M.R., ed. Interaction of ultrasound with biological tissues, Washington, DC, US Government Printing Office, pp. 65-68 (US Dept HEW Publ. (FDA) 73-8008).

DUNN, F. & FRY, F.J. (1971) Ultrasonic threshold dosages for the mammalian central nervous system. IEEE Trans. Biomed. Eng., BME-18: 253-256.

DUNN, F. & POND, J.B. (1978) Selected non-thermal mechanisms of interaction of ultrasound and biological media. In: Fry, F.J., ed. Ultrasound: Its application in medicine and biology, Part II, Amsterdam, Oxford, and New York, Elsevier Sci. Publ. Co., pp. 539-559.

DYER, H.J. (1965) Changes in behaviour of mosses treated with ultrasound. J. Acoust. Soc. Am., 37: 1195A.

DYER, H.J. (1972) Structural effects of ultrasound on the cell. In: Reid, J.M. & Sikov, M.R., ed. Interaction of ultrasound with biological tissues, Washington, DC, pp. 73-75 (US Dept HEW Publ. (FDA) 73-8008).

DYSON, M. & POND, J. (1973) Effects of ultrasound on circulation. Physiotherapy, 59(9): 284-287.

DYSON, M., POND, J.B., JOSEPH, J., & WARWICK, R. (1968) The stimulation of tissue regeneration by means of ultrasound. Clin. Sci., 35: 273-285.

DYSON, M., POND, J.B., & WARWICK, R. (1970) Stimulation of tissue regeneration by pulsed-wave ultrasound. IEEE Trans. Son. Ultrason., SU-17: 133-140.

DYSON, M., WOODWARD, B., & POND, J.B. (1971) Flow of red blood cells stopped by ultrasound. Nature (Lond.), 232: 572-573.

DYSON, M., POND, J.B., WOODWARD, B., & BROADBENT, J. (1974) The production of blood cell stasis and endothelial damage in the blood vessels of chick embryos treated with ultrasound in a stationary wave field. Ultrasound Med. Biol., 1: 133-148.

DYSON, M., FRANKS, C., & SUCKLING, J. (1976) Stimulation of healing of varicose ulcers by ultrasound. Ultrasonics, 14: 232-236.

EDMONDS, P.D. (1972) Effects on macromolecules. In: Reid, J.W. & Sikov, M.R., ed. Interaction of ultrasound and biological tissues, Washington, DC, Government Printing Office, pp. 5-11 (US Dept HEW Publ. (FDA) 73-8008).

EDMONDS, P.D. (1980) Further skeptical comment on reported adverse effects of alleged low intensity ultrasound. In: Proceedings of the 25th American Institute of Ultrasound in Medicine Conference, New Orleans, Washington, DC, American Institute of Ultrasound, p. 50.

EDMONDS, P.D., STOLZENBERG, S.J., TORBIT, C.A., MADAN, S.M., & PRATT, D.E. (1979) Post-partum survival of mice exposed in utero to ultrasound. J. Acoust. Soc. Am., 66(2): 590-593.

ELDER, S.A. (1959) Cavitation microstreaming. J. Acoust. Soc. Am., 31: 54-64.

ELDRIDGE, D.H., Jr (1950) Some responses of the ear to high frequency sound. Am. Soc. Exp. Biol. fed. Proc., 9: 37.

ELMER, W.A. & FLEISCHER, A.C. (1974) Enhancement of DNA synthesis in neonatal mouse tibial epiphyses after exposure to therapeutic ultrasound. J. clin. Ultrasound, 2: 191-195.

EL'PINER, I.E. (1964) Ultrasound - physical, chemical, and biological effects (Translated from Russian, Consultants Bureau, New York).

EMERY, J.M. (1974) Phacoemulsification. In: Emery, J.M. & Paton, D., ed. Proceedings of the Third Biannual Cataract Surgery Congress, St. Louis, Mosby Co., pp. 207-208.

EMERY, J.M. & PATON, D. (1974) Phacoemulsification: A survey of 2875 cases. In: Emery, J.M. & Paton, D., ed. Proceedings of the Third Biannual Cataract Surgery Congress, St. Louis, Mosby Co., pp. 222-224.

EMERY, J.M., LANDIS, D.J., & BENOLKEN, R.M. (1974) The phacoemulsifier: an evaluation of performance. In: Emery, J.M. & Paton, D., ed. Proceedings of the Third Biannual Cataract Surgery Congress, St. Louis, Mosby Co., pp. 208-222.

ESCHE, R. (1952) [Investigation of cavitation by sound in liquids.] Akust. Beihefte, 4: 208-218 (in German).

ESMAT, N. (1975) Investigations of the effects of different doses of ultrasonic waves on the human nerve conduction velocity. J. Egypt. Med. Assoc., 53: 395-402.

ETIENNE, J., FILIPCZYNSKI, L., FIREK, A., GRONIOWSKI, J., LYPACEWICZ, G., & SALKOWSKI, J. (1976) Intensity determination of ultrasonic focused beams used in ultrasonography in the case of gravid uterus. Ultrasound Med. Biol., 2: 119-122.

EVANS, A. & WALDER, D.N. (1970) Detection of circulating bubbles in the intact animal. Ultrasonics, 8: 216-217.

FAHIM, M.S., FAHIM, Z., DER, R., HALL, D.G., & HARMAN, J. (1975) Heat in male contraception (hot water 60 °C, infrared, microwave and ultrasound). Contraception, 11(5): 549-562.

FAHIM, M.S., FAHIM, Z., HARMAN, J., THOMPSON, L., MONTIE, J., & HALL, D.G. (1977) Ultrasound as a new method of male contraception. Fertil. Steril., 28(8): 823-831.

FALLON, J.T., STEPHENS, W.E., & EGGLETON, R.C. (1973) An ultrastructural study of the effect of ultrasound on arterial tissue. J. Pathol., 111: 275-284.

FALUS, M., KORANYI, G., SOBEL, M., PESTI, E., & TRINK, V.B. (1972) Follow-up studies on infants examined by ultrasound during the fetal age. Orvosi Hetilap, 13: 2119-2121.

FARMER, W.C. (1968) Effect of intensity of ultrasound on conduction of motor axons. Phys. Ther., 48(11): 1233-1237.

FARMERY, M.J. & WHITTINGHAM, T.A. (1978) A portable radiation force balance for use with diagnostic ultrasonic equipment. Ultrasound Med. Biol., 3: 373-379.

FINKLER, H. & HAUSLER, E. (1976) Focusing of ultrasonic shockwaves for the touchless destruction of kidney stones. In: Ultrasonics Symposium Proceedings, Annapolis, Maryland, pp. 97-99.

FIRESTONE, F.A. (1945) The supersonic reflectoscope for internal inspection. Met. Prog., 48: 505-512.

FISHMAN, S.S. (1968) Biological effects of ultrasound in vivo and in vitro haemolysis. Proc. West. Pharmacol. Soc., 11: 149-150.

FLYNN, H.G. (1964) Physics of acoustic cavitation in liquids. In: Mason, W.P., ed. Physical acoustics, New York, Academic Press. Vol 1B, pp. 57-172.

FORREST, J.O. (1967) Ultrasonic scaling, a 5-year assessment. Br. dental J., 122: 9-14.

FOSTER, K.R. & WIEDERHOLD, M.L. (1978) Auditory responses in cats produced by pulsed ultrasound. J. Acoust. Soc. Am., 63: 1199-1205.

FRANKLIN, T.D., EGENES, K.M., FALLON, J.T., SANGHVI, N.T., & FRY, F.J. (1977) Therapeutic applications of ultrasound in myocardial infraction: A chronic six-week study in dogs. Presented at: American Institute of Ultrasound in Medicine Meeting, Dallas, Texas, Washington, DC, AIUM.

FREDERIKSEN, E. (1977) Condenser microphones used as sound sources. Brüel Kjaer tech. Rev., 3: 3-32.

FRITZ-NIGGLI, H. & BONI, A. (1950) Biological experiments on Drosophila melanogaster with supersonic vibrations. Science, 112: 120-122.

FROST, H.M. (1977) Heating under ultrasonic dental scaling conditions. In: Symposium on Biological Effects and Characterizations of Ultrasound Sources, Washington, DC, US DHEW, pp. 64-76 (US Dept HEW Publ. (FDA) 78-8048).

FRY, F.J. & BARGER, J.E. (1978) Acoustic properties of the human skull. J. Acoust. Soc. Am., 63: 1576-1590.

FRY, W.J. & DUNN, F. (1956) Ultrasonic irradiation of the central nervous system at high sound levels. J. Acoust. Soc. Am., 28: 129-131.

FRY, F.J., ADES, H.W., & FRY, W.J. (1958) Production of reversible changes in the central nervous system by ultrasound. Science, 127: 83-84.

FRY, F.J., KOSSOFF, G., EGGLETON, R.C., & DUNN, F. (1970) Threshold ultrasonic dosages for structural changes in the mammaliar brain. J. Acoust. Soc. Am., 48: 1416-1417.

FRY, F.J., JOHNSON, L.K., & ERDMANN, W.A. (1978) Interaction of ultrasound with solid tumours in vivo and tumour cell suspensions in vitro. In: White, D. & Lyons, E.A. ed. Ultrasound in medicine, New York, Plenum Press, Vol. 4., pp. 587-588.

FUNG, H.K., CHEUNG, K., LYONS, E.A., & KAY, N.E. (1978) The effects of low-dose ultrasound on human peripheral lymphocyte function in vitro. In: White, D. & Lyons, E.A., ed. Ultrasound in medicine, New York, Plenum Press, Vol. 4, pp. 583-586.

GALPERIN-LEMAITRE, H., GUSTOT, P., & LEVI, S. (1973) Ultrasound and marrow-cell chromosomes. Lancet, 2: 505-506.

GALPERIN-LEMAITRE, H., KIRSCH-VOLDERS, M., & LEVI, S. (1975) Ultrasound and mammalian DNA. Lancet, 2: 662.

GAVRILOV, L.R., NARBUT, N.P., & FRIDMAN, F.E. (1974) [Use of focused ultrasound to accelerate the "mating" of a cataract.] Akustič. Ž. (USSR), 20: 274-377 (in Russian).

GAVRILOV, L.R., TSIRULNIKOV, E.M., & SHCHEKANOV, E.E. (1975) [Stimulation of auditory receptors by focused ultrasound.] Sov. Phys. Acoust., 21(5): 437-489 (in Russian).

GAVRILOV, L.R., GERSUNI, G.V., ILYINSKI, O.B., SIROTYUK, M.G., TSIRULNIKOV, E., & SHCHEKANOV, E.E. (1976) The effect of focused ultrasound on the skin and deep nerve structures of man and animal. Brain Res., 43: 279-292.

GAVRILOV, L.R., GERSUNI, G.V., ILYINSKI, O.B., TSIRULNIKOV, E.M., & SHCHEKANOV, E.E. (1977) A study of reception with the use of focused ultrasound - effects on the skin and deep receptor structures in man. Brain Res., 135(2): 265-277.

GERASIMOVA, E.J. (1976) [A study of the effect of ultrasound on the sympathicoadrenal system of workers.] Gig. i Sanit., 8: 23-29 (in Russian).

GERNER, E.W., BOONE, R., CONNER, W.G., HICKS, J.A., & BOONE, M.M. (1976) A transient thermotolerant survival response produced by single thermal doses in Hela cells. Cancer Res., 36: 1035-1040.

GERSHOY, A. & NYBORG, W.L. (1973) Perturbation of plant-cell contents by ultrasonic microirradiation. J. Acoust. Soc. Am., 54(5): 1356-1367.

GERSTEN, J.W. (1955) Effect of ultrasound on tendon extensibility. Am. J. phys. Med., 34: 362-369.

GIRARD, L.J. (1974) Ultrasonic aspiration - irrigation of cataract and the vitreous. In: Emery, J.M. & Paton, D.J., ed. Proceedings of the Third Biannual Cataract Surgical Congress. Current concepts in cataract surgery, St. Louis, Mosby Co., pp. 194-197.

GLICK, D., ADAMOVICS, A., EDMONDS, P.D., & TAENZER, J.C. (1979) Search for biochemical effects in cells and tissues of ultrasonic irradiation of mice and of the in vitro irradiation of mouse peritoneal and human amniotic cells. Ultrasound Med. Biol., 5: 23-33.

GLICK, D., NOLAN, H.W., & EDMONDS, P.D. (1981) Blood chemical and haematological effects of ultrasonic irradiation of mice. Ultrasound Med. Biol., 7: 87-90.

GLOERSON, W.R., HARRIS, G.R., STEWART, H.F., & LEWIN, P.A. (1982) A comparison of two calibration methods for ultrasonic hydrophones. Ultrasound Med. Biol., 8: 545-548.

GLOVER, C.J., McINTIRE, L.V., LEVERETT, L.B., HELLUMS, J.D., BROWN, C.H., & NATELSON, E.A. (1974) Effect of shear stress on clot structure formation. Trans. Am. Soc. Artif. Int. Org., 20: 463-468.

GOLDBLAT, V.I. (1969) [Processes of bone tissue regeneration under the effect of ultrasound.] Ortoped. Travmoto. Proteziro., 30: 57-61 (in Russian).

GOLDSTEIN, A. (1982) Quality assurance in diagnostic ultrasound. In: Repacholi, M.H. & Benwell, D.A., ed. Essentials of medical ultrasound, New Jersey, Humana Press, pp. 215-280.

GOLIAMINA, L.P. (1974) Ultrasonic surgery. In: Proceedings of the 8th International Congress on Acoustics, Guildford UK, IPC Science and Technology Press, pp. 63-69.

GORALČUK, M.V. & KOŠIK, T.F. (1976) [The effects of ultrasound on histological and histochemical changes in the healing process of suppurative ulcers of the cornea.] Ofthalmol. Ž., 31(7): 533-535 (in Russian).

GORSLIKOV, S.I., GOREUNOV, O.N., & ANTROPOV, S.A. (1965) Biological effects of ultrasound. Ultrasonics, 4: 211.

GOSS, S.A., COBB, J.W., & FRIZZELL, L.A. (1977) Effect of beam width and thermocouple size on the measurement of ultrasonic absorption using thermoelectric technique. In: 1977 Ultrasonic Symposium Proceedings, New York, IEEE, pp. 206-211.

GREGORY, W.D., MILLER, M.W., CARSTENSEN, E.L., CATALDO, F.L., & REDDY, M.M. (1974) Nonthermal effects of 2 MHz ultrasound on the growth and cytology of Vicia faba roots. Br. J. Radiol., 47: 122-129.

GRIGOR'EVA, V.M. (1966a) Effect of ultrasonic vibrations on personnel working with ultrasonic equipment. Sov. Phys. Acoust., 11: 426-427.

GRIGOR'EVA, V.M. (1966b) [Ultrasound and the question of occupational hazards.] Mascinstreočija, 8: 32 (in Russian) (Abstract in Ultrasonics, 4: 214).

HAHN, G.M., BRAUN, J., & HAR-KEDAR, I. (1975) Thermo-chemotherapy: synergism between hyperthermia (42–43 degrees) and adriamycin (or bleomycin) in mammalian cells inactivation. Proc. Natl Acad. Sci. USA, 72: 937–940.

HARA, K. (1980) Effect of ultrasonic irradiation on chromosomes, cell division and developing embryos. Acta Obst. Gynaecol. Jpn , 32(1): 61–68.

HARA, K., MINOURA, S., OKAI, T., & SAKAMOTO, S. (1977) Symposium on recent studies in the safety of diagnostic ultrasound. Safety of ultrasonics on organism. Jpn. J. med. Ultrasonics, 4: 256–258.

HARRIS, G.R. (1981) Detection and analysis of transient ultrasonic fields: A study using polyvinylidene fluoride piezoelectric polymer hydrophones. PhD Thesis, Catholic University of America, Washington, DC.

HARRIS, G.R., HERMAN, B.A., HARAN, M.E., & SMITH, S.W. (1977) Calibration and use of miniature ultrasonic hydrophones. In: Symposium on Biological Effects and Characterization of Ultrasound Sources, Wasington, DC, US DHEW, pp. 169–174 (US Dept HEW Publ. (FDA) 78–8048).

HARVEY, W., DYSON, M., POND, J.B., & GRAHAME, R. (1975) The in vitro stimulation of protein synthesis in human fibroblasts by therapeutic levels of ultrasound. In: Kazner, E. et al., ed. Proceedings of the 2nd European Congress on Ultrasonics in Medicine, Munich 12–16 May 1975, Amsterdam, Excerpta Medica, pp. 10–21. (Excerpta Medica International Congress Series No. 363).

HAUPT, M., MARTIN, A.O., SIMPSON, J.L., IQBAL, M.A., ELIAS, S., & SABBAGHA, R.E. (1981) Ultrasonic induction of sister chromatid exchanges in human lymphocytes. Human Genet., 59: 221–226.

HELLMAN, L.M., DUFFUS, G.M., DONALD, L., & SUNDEN, B. (1970) Safety of diagnostic ultrasound in obstetrics. Lancet, 1: 1133–1135.

HEIMBURGER, R.F., FRY, F.J., FRANKLIN, T.D., & EGGLESTON, R.C. (1975) Ultrasound potentiation of chemotherapy for brain malignancy. In: White, D., ed. Ultrasound in medicine, New York, Plenum Press, Vol 1., p. 273.

HERMAN, B.A. & POWELL, D. (1981) Airborne ultrasound: Measurement and possible adverse effects, Washington, DC, (US Dept Health and Human Services, HHS Publ. (FDA) 81-8163).

HERTZ, R.H., TIMOR TRITSCH, I., DIERKER, L.J., CHIK, L., & ROSEN, M.G. (1979) Continuous ultrasound and fetal movement. Am. J. Obstet. Gynecol., 135(1): 152-154.

HILL, C.R. (1971) Acoustic intensity measurement on ultrasonic diagnostic devices. In: Bock, J. & Ossoinig, K., ed. Ultrasonographia medica, Vienna, Vienna Academy of Medicine, pp. 21-27.

HILL, C.R. (1972a) Ultrasonic exposure thresholds for changes in cells and tissues. J. Acoust. Soc. Am., 52: 667-672.

HILL, C.R. (1972b) Interaction of ultrasound with cells. In: Reid, J.M. & Sikov, M.R., ed. Interaction of ultrasound with biological tissues, Washington, DC, US DHEW, pp. 57-79 (US Dept HEW Publ. (FDA) 73-8008).

HILL, C.R. & JOSHI, G.P. (1970) The significance of cavitation in interpreting the biological effects of ultrasound. In: Proceedings of a Conference on Ultrasonics in Biology and Medicine, Warsaw, UBIOMED-70, pp. 125-131.

HILL, C.R. & TER HAAR, G. (1981) Ultrasound. In: Suess, M.J., ed. Nonionizing radiation protection, Copenhagen, World Health Organization Regional Office for Europe (WHO Regional Publications, European Series No. 10).

HILL, C.R., CLARKE, P.R., CROWE, M.R., & HAMMICK, J.W. (1969) Biophysical effects of cavitation in an 1 MHz ultrasonic beam. In: Ultrasonics for Industry Conference Papers, 1969, pp. 26-30.

HILL, C.R., JOSHI, G.P., & REVELL, S.H. (1972) A search for chromosome damage following exposure of Chinese hamster cells to high intensity, pulsed ultrasound. Br. J. Radiol., 45: 333-334.

HODGSON, W.J.B., BAKARE, S., HARRINGTON, E., FINKELSTEIN, J., PODDAR, P.K., LOSCALZO, L.L., WEITZ, J., & McELHINNEY, A.J. (1979) General surgical evaluation of a powered device operating at ultrasonic frequencies. Mt Sinai J. Med. (NY), 46(2): 99-103.

HOUNSFIELD, G.N. (1973) Computerized traverse axial scanning (tomography), Part I. Description of system. Br. J. Radiol., 46: 1016-1022.

HRAZDIRA, I. & ADLER, J. (1980) Electrokinetic properties of isolated cells exposed to low levels of ultrasound. In: Ultrasound Interactions in Biology in Medicine, International Symposium, Nov. 10-14, Casel Reinhardsbrunn-GDR, p. C-11.

HRAZDIRA, I. & HAVELKOVA, M. (1966) Ultrasound and the ultramicroscopic structure of Rhizopus nigricans. Naturwissenschaften, 53: 206.

HRAZDIRA, I. & KONECNY, M. (1966) Functional and morphological changes in the thyroid gland after ultrasonic irradiation. Am. J. Phys. Med., 45(5): 238-243.

HU, J.H. & ULRICH, W.D. (1976) Effects of low-intensity ultrasound on the central nervous system of primates. Aviat. Space environ. Med., 47(6): 640-643.

HU, J.H., TAYLOR, J.D., PRESS, H.C., & WHITE, J.E. (1978) Ultrasonic effects on mammalian interstitial muscle membrane. Aviat. Space environ. Med., 49(4): 607-609.

HUETER, T.F. & BOLT, R.H. (1955) Sonics. In: Radiation pressure, New York, Wiley, pp. 43-53.

HUG, O. & PAPE, R. (1954) Establishing the presence of ultrasound cavitation in tissues. Stralentherapie, 94: 79-99 (translated from German).

HUSTLER, J.E., ZAROD, A.P., & WILLIAMS, A.R. (1978) Ultrasonic modification of experimental bruising in the guinea-pig pinna. Ultrasonics, 16(5): 223-228

IDE, M. & OHIRA, E. (1975) Measurement of ultrasonic noise radiated from ultrasonic cleaners. In: Proceedings of the Acoustical Society of Japan, pp. 135-136

IERNETTI, G. (1971) Cavitation threshold dependence on volume. Acustica, 24: 191-196.

IEC (1980a) Draft: IEC Standard Publication 601-2-XX. Ultrasonic Medical Diagnostic Equipment, Part 2. Particular requirements for safety (IEC/TC 62D (Sec) 31, Dec. 1980).

IEC (1980b) Draft: Ultrasonic therapy equipment, particular requirements for safety (IEC/TC 62/SC 62D (Sec.) 25, Dec. 1980).

IEC (1981) Draft: Characteristics and calibration of hydrophones for operation in the frequency range 0.5 to 15 MHz (IEC/TC 29/SC 29D (Central Office) 19).

IEC (1982) Draft: Methods of measuring the performance of ultrasonic pulse-echo diagnostic equipment (IEC/TC 29/SC 29D (Central Office) 16, February 1982).

IKEUCHI, T., SASAKI, M., OSHIMURA, M., AZUMI, J., TSUJI, K., & SHIMIZU, T. (1973) Ultrasound and embryonic chromosomes. Br. med. J., 1: 112.

ILO (1977) Protection of workers against noise and vibration in the working environment, Geneva, International Labour Organization, pp. 66 (ILO Codes of Practice).

IRPA (1977) Overviews on non-ionizing radiation. Washington, DC, International Radiation Protection Association, US Dept of Health, Education and Welfare, pp. 42-59.

IRPA (1981) Draft: Guidelines on limits of human exposure to airborne acoustic energy having one-third octave bands with mid frequencies from 8 to 50 kHz. International Radiation Protection Association, International Non-Ionizing Radiation Committee (IRPA/INIRC), Nov. 1981.

JACKE, S.E. (1979) Ultrasonics in industry today. In: Proceedings Ultrasonics International 1979, Graz, Austria, Guildford, UK, IPC Science and Technology Press.

JACOBSON, E.J., DOWNS, M.P., & FLETCHER; J.L. (1969) Clinical findings in high frequency thresholds during known drug usage. J. aud. Res., 9: 379.

JAMES, J.A. (1963) New developments in ultrasonic therapy of Ménière's disease. Ann. R. Coll. Surg. Engl., 33: 226-244.

JANKOWIAK, J. & MAJEWSKI, C. (1966) Electron-microscope studies of acid phosphates in neutrophilic granulocytes in the blood of rabbits subjected to ultrasound. Am. J. phys. Med., 45(1): 1-7.

JAPANESE MINISTRY OF LABOUR (1971) Airborne ultrasound standard, order by Chief of the Labour Standard Bureau based on the Circular 326 of the Japanese Ministry of Labour. Guidelines on the use of ultrasonic welder, Tokyo, Japan.

JIS (1979) Japanese Industrial Standards. Ultrasonic Doppler fetal diagnostic equipment (draft, March 1979). Tokyo, Japan, Electronic Industries Association of Japan.

JIS (1980) Japanese Industrial Standards, Draft: M-mode ultrasonic diagnostic equipment. Tokyo, Japan.

JIS (1981) Japanese Industrial Standards. Draft: Methods of measuring the performance of ultrasonic pulse-echo diagnostic equipment. Tokyo, Japan.

JOHNSON, A. & LINDVALL, A. (1969) Effects of low-intensity ultrasound in viscous properties of Elodea cells. Naturwissenschaft, 56: 40.

JOHNSON, W.N. & WILSON, J.R. (1957) The application of the ultrasonic dental unit to scaling procedures. J. Periodontol., 23: 264-271.

JOSHI, G.P., HILL, C.R., & FORRESTER, J.A. (1973) Mode of action of ultrasound on the surface change of mammalian cells in vitro. Ultrasound Med. Biol., 1: 45-48.

JSA (1976) Japanese Standards Association, Draft: A-mode ultrasonic diagnostic equipment. Tokyo, Japan.

JSA (1978) Japanese Standards Association, Draft: Japanese Industrial Standard, Manual scanning B-mode ultrasonic diagnostic equipment, March, Tokyo, Japan.

KARDUCK, A. & WEHMER, W. (1974) Morphologic studies of the influence of ultrasound upon the growing rabbit's larynx. Arch. Oto-Rhinol.-Laryngol., 206: 137-154.

KATO, M. (1969) Ultrasonic effects affecting the mechanism of reproduction of micronsized microorganism. J. Phys. Soc. Jpn, 31: 31-32.

KAUFMANN, J.S. & KREMKAU, F.W. (1978) Influence of ultrasound on mouse leukaemia cell CNA synthesis, membrane integrity, and uptake of anticancer drugs in vitro. In: White, D. & Lyons, E.A., ed. Ultrasound in medicine, New York, Plenum Press, Vol. 4, pp. 589-590.

KAUFMAN, G.E. & MILLER, M.W. (1978) Growth retardation in Chinese hamster V-79 cells exposed to 1 MHz ultrasound. Ultrasound Med. Biol., 4: 139-144.

KAUFMAN, G.E., MILLER, M.W., GRIFFITHS, T.D., CIARAVINO, V., &
CARSTENSEN, E.L. (1977) Lysis and viability of cultured
mammalian cells exposed to 1 MHz ultrasound. Ultrasound Med.
Biol., 3: 21-25.

KELMAN, C.D. (1967) Phaco-emulsification and aspiration: A
new technique of cataract removal. Am. J. Ophthalmol., 64:
23-35.

KHOE, W.H. (1977) Ultrasound acupuncture: effective
treatment modality for various diseases. Am. J. Acupunct.,
5(1): 31-34.

KINSLER, L.E. & FREY, P. (1962) Fundamentals of acoustics,
New York, J. Wiley Press.

KISHI, M., MISHIMA, T., ITAKURA, T., TSUDA, K., & OKA, M.
(1975) Experimental studies of effects of intense ultrasound
on implantable murine glioma. In: Kazner, E., de Vliger, M.,
Muller, H.R., & McCready, V.R., ed. Ultrasonics in medicine,
Amsterdam, Excerpta Medica, pp. 28-33.

KLEINSCHMIDT, P. & MAGORI, V. (1981) Ultrasonic remote
sensors for noncontact object detection. Siemans Forsch-u.
Entwickl.-Ber., 10(2): 110-118.

KNIGHT, J.J. (1968) Effects of airborne ultrasound on man.
Ultrasonics, 6: 39-42.

KNIGHT, J.J. & COLES, R.R.A. (1966) A six-year prospective
study of the effect of jet aircraft noise on hearing. J. R.
Nav. Med. Serv., 52: 92.

KOH, S. (1981) The safety of diagnostic continuous wave
ultrasonic irradiation - a clinical study. Serum hemoglobin
level and scanning electron microscopic finding of maternal
and cord blood in vitro. Acta Obstet. Gynaec. Jpn., 33:
469-478.

KOIFMAN, M.M., VACILIEVA, T.N., MASLOV, K.I., MAEV, R.G., &
LEVIN, V.M. (1980) Antibody secretion changes induced by
ultrasound in lymphoid cells. In: Ultrasound Interaction in
Biology and Medicine. International Symposium, Nov. 10-14,
1980. Castle Reinhardsbrunn-GDR, p. C-16.

KOLAR, J., BABICKJ, A., KASLOVA, J., & KASI, J. (1965) [The
effect of ultrasound on the mineral metabolism of bones.]
Travmatol. protezinov., 26(8): 43-51 (in Russian).

KOSSOFF, G. (1978) On the measurement and specification of acoustic output generated by pulse ultrasonic diagnostic equipment. J. clin. Ultrasound, 6(5): 303-309.

KOSSOFF, G. & KHAN, A.E. (1966) Treatment of vertigo using the ultrasonic generator. Arch. Otolaryngol., 84: 181-188.

KREMKAU, F.W. (1979) Cancer therapy with ultrasound: a historical review. J. clin. Ultrasound, 7: 287-300.

KREMKAU, F.W. & CARSTENSEN, E.L. (1972) Macromolecular interaction in sound absorption. In: Reid, J.M. & Sikov, M.R., ed. Interaction of ultrasound and biological tissues, Washington, DC, US DHEW, pp. 37-42 (US Dept HEW Publ. (FDA) 73-8008).

KREMKAU, F.W. & WITCOFSKI, R.L. (1974) Mitotic reduction in rat liver exposed to ultrasound. J. clin. Ultrasound, 2(2): 123-126.

KUNZE-MUHL, E. (1981) Observation of the effect of X-rays alone and in combination with ultrasound on human chromosomes. Human Genet., 57: 257-260.

KURACHI, K., CHIBA, Y., SUEHARA, N., & SAKUMOTO, T. (1981) Studies on the effect of pulsed ultrasound on chromosome and erythrocyte, and optimal utility of ultrasound diagnosis in early pregnancy. Jpn. J. med. Ultrason., 8: 271-273.

LATT, S.A. & SCHRECK, R.R. (1980) Sister chromatid exchange analysis. Am. J. Hum. Genet., 32(3): 297-313.

LEHMANN, J.F. (1965a) Ultrasonic diathermy. In: Krusen, F.H., Kottke, F.J., & Ellerwood, P., ed. Handbook of physical medicine and rehabilitation, Philadelphia and London, W.B. Saunders Co., pp. 271-299.

LEHMANN, J.F. (1965b) Ultrasound and therapy. In: Licht, E. & Kamenetz, H.L., ed. Therapeutic heat and cold. 2nd ed., Baltimore, Maryland, Waverley Press Inc., pp. 321-386.

LEHMANN, J.F. & GUY, A.W. (1972) Ultrasound therapy. In: Reid, J.M. & Sikov, M.R., ed. Interaction of ultrasound and biological tissues, Washington, DC, US DHEW, pp. 141-152. (HEW Publ. (FDA) 73-8008).

LEHMANN, J.F. & HERRICK, J.F. (1953) Biologic reactions to cavitation, a consideration for ultrasonic therapy. Arch. Phys. Med., 34: 86-98.

LEHMANN, J.F., McMILLAN, J.A., BRUNNER, G.D., & BLUMBERG, J.B. (1959) Comparative study of the efficiency of shortwave, microwave and ultrasonic diathermy in heating the hip joint. Arch. Phys. Med., 40: 510-512.

LEHMANN, J.F., DeLATEUR, B.J., STONEBRIDGE, J.B., & WARREN, C.G. (1967) Therapeutic temperature distribution produced by ultrasound as modified by dosage and volume of tissue exposed. Arch. Phys. Med., 48(12): 662-666.

LEHMANN, J.F., WARREN, C.G., & SCHAM, S.M. (1974) Therapeutic heat and cold. In: Urist, M.R., ed. Clinical orthopaedics and related research, Toronto, Lippincott Company, pp. 207-245.

LEHMANN, J.F., WARREN, C.G., & GUY, A.W. (1978) Therapy with continuing wave ultrasound. In: Fry, F.J., ed. Ultrasound: Its application in medicine and biology, Amsterdam, Elsevier Press, pp. 561-587.

LELE, P.P. (1967) Production of deep focal lesions by focused ultrasound - current status. Ultrasonics, 5: 105-112.

LELE, P.P. (1975) Ultrasonic teratology in mice and man. In: Proceedings of the Second European Congress of Ultrasonics in Medicine, Munich, 12-16 May, Amsterdam, Excerpta Medica, pp. 22-27.

LELE, P.P. & PIERCE, A.D. (1972) The thermal hypothesis of the mechanism of ultrasonic focal destruction in organized tissues. In: Interaction of ultrasound and biological tissues, Washington, DC, US DHEW, pp. 121-128 (HEW Publ. (FDA) 73-8008).

LEMONS, R.A. & QUATE, C.F. (1975) Acoustic microscopy - a tool for medical and biological research, New York, Plenum Press, pp. 305-317.

LERNER, R., CARSTENSEN, E., & DUNN, F. (1973) Frequency dependence of thresholds for ultrasonic production of thermal lesions in tissue. J. Acoust. Soc. Am., 54: 504-506.

LEVERETT, L.B., HELLUMS, J.D., ALFREY, C.P., & LYNCH, E.C. (1972) Red blood cell damage by shear stress. Biophys. J., 12: 257-273.

LEWIN, P.A. (1978) Ultrasound-induced damage of biological tissue. PhD Thesis, AFM 78-16, Copenhagen, Technical University of Denmark.

LEWIN, P.A. (1981a) Calibration and performance evaluation of miniature ultrasonic hydrophone using time delay spectrometry. In: Proceedings of the IEEE Ultrasonics Symposium, October 1981, pp. 660-664.

LEWIN, P.A. (1981b) Miniature piezoelectric polymer ultrasonic hydrophone probes. Ultrasonics, 19: 213-216.

LEWIN, P.A. & CHIVERS, R.C. (1980) On viscoelastic models of the cell membrane. Acoust. Lett., 4(5): 85-89.

LI, G.C., HAHN, G.M., & TOLMACH, L.J. (1977) Cellular inactivation by ultrasound. Nature (Lond.), 267: 163-165.

LIEBESKIND, D., BASES, R., ELEQUIN, F., NEUBORT, S., LEIFER, R., GOLDBERG, R., & KOENIGSBERG, M. (1979a) Diagnostic ultrasound: effects on the DNA and growth patterns of animal cells. Radiology, 131: 177-184.

LIEBESKIND, D., BASES, R., MENDEZ, F., ELEQUIN, F., & KOENIGSBERG, M. (1979b) Sister chromatid exchanges in human lymphocytes after exposure to diagnostic ultrasound. Science, 205: 1273-1275.

LIEBESKIND, D., BASES, R., KOENIGSBERG, M., KOSS, L., & RAVENTOS, C. (1981a) Morphological changes in the surface characteristics of cultured cells after exposure to diagnostic ultrasound. Radiology, 138: 419-423.

LIEBESKIND, D., PADAWER, J., WOLLEY, R., & BASES, R. (1981b) Diagnostic ultrasound: Time-lapse and transmission electron microscopic studies of cells insonated in vitro. Presented at the L.H. Gray Conference in Oxford, England, July 13-16. New York, Albert Einstein College of Medicine.

LINDSTRÖM, K. & SVEDMAN, P. (1981) Ultrasound real-time scanner used in air for imaging objects in the ambient environment. IRCS med. Sci. biomed. Technol., 9: 132.

LINDSTRÖM, K. MAURITZSON, L., BENONI, G., SVEDMAN, P., & WILLNER, S. (1982) Application of air-borne ultrasound to biomedical measurements. Med. biol. Eng. Comput., 20: 392-400.

LIZZI, F.L., COLEMAN, D.J., DRILLER, J., FRANZEN, L.A., & JAKOBIEC, F.A. (1978a) Experimental ultrasonically induced lesions in the retina, choroid and sclera. Invest. Ophthalmol. Visual Sci., 17(4): 350-360.

LIZZI, F.L., PACKER, A.J., & COLEMAN, D.J. (1978b)
Experimental cataract production by high frequency ultrasound.
Ann. Ophthalmol., 10: 934-942.

LONGO, F.W., TOMASHEFSKY, P., RIVEN, B.D., LONGO, W.E.,
LATTIMER, J.K., & TENNENBAUM, M. (1979) Interaction of
ultrasound with neoplastic tissue. Local effect on
subcutaneously implanted Furth-Columbia rat Wilm's tumor.
Urology, VI: 631-634.

LOTA, M.J. & DARLING, R.C. (1955) Changes in permeability of
red blood cell membrane in a homogeneous ultrasonic field.
Arch. phys. Med. Rehabil., 36: 282-287.

LOVE, L.A. & KREMKAU, F.W. (1980) Intracellular temperature
distribution produced by ultrasound. J. Acoust. Soc. Am., 67:
1045-1050.

LUK, K.H., HULSE, R.M., & PHILLIPS, T.L. (1980) Hyperthermia
in cancer therapy. Western J. Med., 132: 179-185.

LUNAN, K.D., WEN, A.C., BARFOD, E.T., EDMONDS, P.D., & PRATT,
D.E. (1979) Decreased aggregation of mouse platelets after
in vivo exposures to ultrasound. Thromb. Haemos., 40: 568-570.

LYNNWORTH, L.C. (1975) Industrial applications of ultrasound
- a review. II. Measurements, tests and process control using
low intensity ultrasound. IEEE Trans. Son. Ultrason.,
SU-22(2): 71-101.

LYON, M.F. & SIMPSON, G.W. (1974) An investigation into the
possible genetic hazards of ultrasound. Br. J. Radiol., 47:
712-722.

LYONS, E.A. (1982) Clinical applications of diagnostic
ultrasound. In: Repacholi, M.H. & Benwell, D.A., ed.
Essentials of medical ultrasound, New Jersey, Humana Press,
pp. 141-180.

LYONS, E.A. & COGGRAVES, M. (1979) Follow-up study in
children exposed to ultrasound in utero - an interim report.
Abstract. American Institute of Ultrasound in Medicine
Meeting, Montreal.

MacINTOSH, I.J.C. & DAVEY, D.A. (1970) Chromosome
aberrations induced by an ultrasonic fetal pulse detector. Br.
med. J., 4: 92-93.

MacINTOSH, I.J.C. & DAVEY, D.A. (1972) Relationship between intensity of ultrasound and induction of chromosome aberrations. Br. J. Radiol., 45: 320-327.

MacINTOSH, I.J.C., BROWN, R.C., & COAKLEY, W.T. (1975) Ultrasound and in vitro chromosome aberrations. Br. J. Radiol., 48: 230-232.

McCLAIN, R.M., HOAR, R.M., & SALTZMAN, M.B. (1972) Teratologic study of rats exposed to ultrasound. Am. J. Obstet. Gynecol., 114: 39-42.

MAEDA, K. & MURAO, F. (1977) Studies on the influence of ultrasound irradiation on the growth of cultured cell in vitro. In: White, D. & Brown, R.E., ed. Ultrasound in medicine, New York, Plenum Press, Vol 3B, pp. 2045-2049.

MAJEWSKI, C., KALINOWSKI, M., & JANKOWIAK, J. (1966) Electron-microscopic studies of acid phosphatase activity in the liver of rats subjected to ultrasound. Am. J. phys. Med., 45(5): 234-237.

MANLEY, D.M.J.P. (1969) Ultrasonic detection of gas bubbles in blood. Ultrasonics, 7: 102-105.

MARMOR, J.B., MILERIC, F.J., & HAHN, G.M. (1979) Tumour eradication and cell survival after localized hyperthermia induced by ultrasound. Cancer Res., 39: 2166-2171.

MARMUR, R.K. & PLEVINSKIS, V.P. (1978) [Ultrasonic effects of various intensities on manifestations and duration of poststimulatory cytochemical changes in the retina.] Oftalmol. Ž., 33(4): 287-290 (in Russian).

MARTIN, C.J., GIMMELL, H.G., & WATMOUGH, D.J. (1978) A study of streaming in plant tissue induced by a Doppler fetal heart detector. Ultrasound Med. Biol., 4: 131-138.

MARTIN, C.J., GREGORY, D.W., & HODGEKISS, M. (1981) The effects of ultrasound in vivo on mouse liver in contact with an aqueous coupling medium. Ultrasound Med. Biol., 7(3): 253-265.

MARTINS, B.I. (1971) A study of the effects of ultrasonic waves on reproductive integrity of mammalian cells cultures in vitro. PhD Thesis, University of California (AEC Contract No. W-7405-ang-48 Publ. LBL-37).

MASON, W.P. (1976) Sonics and ultrasonics: early history and applications. In: Ultrasonics Symposium Proceedings, pp. 610-617.

MERINO, O.R., PETERS, L.J., MASON, K.A., & WITHERS, H.R. (1978) The effect of hyperthermia on the radiation response of mouse jejunum. Int. J. Radiat. Oncol. Biol. Phys., 4: 407-414.

MICHAEL, P.L., HERMAN, R.L., BIENVENUE, G.R., & PROUT, H. (1974) An evaluation of industrial acoustic radiation above 10 kHz, Washington, DC, US DHEW (US Dept HEW Publ. (HSM) 99-72-125).

MILLER, D.L., NYBORG, W.L., & WHITCOMB, C.C. (1979) Platelet aggregation induced by ultrasound under specialized conditions in vitro. Science, 205(3): 505-507.

MILLER, J.C., LEITH, J.T., VEOMETT, R.C., & GERNER, E.W. (1976b) Potentiation of radiation myelitis in rats by hyperthermia. Br. J. Radiol., 49: 895-896.

MILLER, M.W., KAUFMAN, G.E., CATALDO, F.L., & CARSTENSEN, E.L. (1976a) Absence of mitotic reduction in regenerating rat liver exposed to ultrasound. J. clin. Ultrasound, 4: 169-172.

MILLER, M.W., CIARAVINO, V., & KAUFMAN, G.E. (1977) Colony size and giant cell formation from mammalian cells exposed to 1 MHz ultrasound radiation. Radiat. Res., 71: 628-634.

MILLER, W.F., JOHNSTON, F.F., & TARKOFF, M.P. (1968) Use of ultrasonic aerosols with ventilatory assisters. J. Asthma Res., 5: 335-354.

MOISEEVA, N.N. & GAVRILOV, L.R. (1977) [The influence of focused high frequency ultrasound on eye tissues.] Oftalmol. V., 32(3): 610-613 (in Russian).

MOLINARI, G.A. (1968a) [Low intensity ultrasound irradiation of the cochlea through the round window: 1. Changes of microphonic potentials]. Bull. Soc. Ital. Biol. Sper., 44: 403-406 (Canadian Govt. Trans. from Italian).

MOLINARI, G.A. (1968b) Low intensity ultrasound irradiation of the cochlea through the round window: II. Changes in the action potentials. Boll. Soc. Ital. Biol. Sper., 44: 406-408 (in Italian).

MÖLLER, P. & GREVSTAD, A.O., & KRISTOFFERSEN, T. (1976) Ultrasonic scaling of maxillary teeth causing tinnitus and temporary hearing shifts. J. clin. Peridontol., 3: 123-127.

MOORE, J.L. & COAKLEY, W.T. (1977) Ultrasonic treatment of Chinese hamster cells at high intensities and long exposure times. Br. J. Radiol., 50: 46-50.

MOORE, R.M., BARRICK, M.K., & HAMILTON, P.M. (1982) Effects of sonic radiation on growth and development. Am. J. Epidemiol., 116(3): 571 (abstract).

MOROHASHI, T. & IIZUKA, R. (1977) Symposium on recent studies in the safety of diagnostic ultrasound. The development of low power ultrasonic instruments. Jpn. J. med. Ultrasound, 4: 271-273.

MORRIS, S.M., PALMER, C.G., FRY, F.J., & JOHNSON, L.K. (1978) Effect of ultrasound on human leucocytes. Sister chromatid exchange analysis. Ultrasound Med. Biol., 4: 253-258.

MORTIMER, A.J., ROY, O.Z., TAICHMAN, G.C., KEON, W.J., & TROLLOPE, B.J. (1978) The effects of ultrasound on the mechanical properties of rat cardiac muscle. Ultrasonics, 16(4): 179-182.

MOSKOW, B. & BRESSMAN, B. (1964) Cemental response to ultrasonic and hand instrumentation. J. Am. Dent. Assoc., 68: 698-703.

MUIR, T.G. & CARSTENSEN, E.L. (1980) Prediction of non-linear acoustic effects at biomedical frequencies and intensities. Ultrasound Med. Biol., 6: 345-357.

MUKUBOH, M., OKAI, T., UEZUMA, S., BABA, K., MINOURA, S., KUMAGAI, K., HARA, K., & SAKAMOTO, S. (1981) [The safety and irradiation effect of ultrasonic real time scanner on fetal development.] Nippon Choompa Igakkai Koen-Rombunshu, 38: 555-556 (in Japanese).

MUMMERY, C.L. (1978) Effect of ultrasound on fibroblasts in vitro. PhD Thesis, University of London.

MURAI, N., HOSHI, K., & NAKAMURA, T. (1975a) Effects of diagnostic ultrasound irradiated during fetal stage of development on orienting behaviour and reflex ontogeny in rats. Tohoku J. exp. Med., 116: 17-24.

MURAI, N., HOSHI, K., KANG, C.-H., & SUZUKI, M. (1975b) Effects of diagnostic ultrasound irradiated during fetal stage on emotional and cognitive behaviour in rats. Tohoku, J. exp. Med., 117: 225-235.

NATIONAL BUREAU OF STANDARDS (USA) (1973) Developing ultrasound standards for use in medicine, industry, research. Noise Control Rep., 2(15): 148.

NEPPIRAS, E.A. (1980) Acoustic cavitation threshold and cyclic processes. Ultrasonics, 18(5): 201-209.

NEVARIL, C.G., LYNCH, E.C., ALFREY, C.R. Jr, & HELLUMS, J.D. (1968) Erythrocyte damage and destruction induced by shear stress. J. lab. clin. Med., 71: 784-790.

NORTHERN, J.L., DOWN, M.P., RUDMOSE, W., GLORIG, A., & FLETCHER, J.L. (1962) Recommended high frequency audiometric threshold levels (8000-18 000 Hz). J. Acoust. Soc. Am., 52: 585-595.

NYBORG, W.L. (1977) Physical mechanisms for biological effects ultrasound, Washington DC, US DHEW (US Dept. HEW Pub. (FDA), 78-8062).

NYBORG, W.L. (1978) Physical principles of ultrasound. In: Fry, F.J., ed. Methods and phenomena 3, Ultrasound: Its applications in medicine and biology, Part I, Amsterdam, Elsevier Scientific Publishing Co., pp.1-75.

NYBORG, W.L. (1979) Physical mechanisms for biological effects of ultrasound. In: Repacholi, M.H. & Benwell, D.A., ed. Ultrasound short course transactions 1979, Health and Welfare, Canada, pp. 83-126.

NYBORG, W.L. (1982) Biophysical mechanisms of ultrasound. In: Repacholi, M.H. & Benwell, D.A., ed. Essentials of medical ultrasound, New Jersey, Humana Press, pp. 35-75.

NYBORG, W.L. & DYER, H.W. (1960) Ultrasonically induced motions in single plant cells. In: Proceedings of the 2nd International Conference on Medical Electronics, pp. 391-396.

NYBORG, W.L., MILLER, D.L., & GERSHOY, A. (1975) Physical consequences of ultrasound on plant tissues and other bio-systems. In: Michaelson, S.M., Miller, M.W., Magin, R., & Carstensen, E.L., ed. Fundamental and applied aspects of non-ionizing radiation, New York and London, Plenum Press, pp. 277-299.

O'BRIEN, W.D. Jr (1976) Ultrasonically induced fetal weight reduction in mice. In: Ultrasound in medicine, New York, Plenum Press, pp. 531-532.

O'BRIEN, W.D. (1978) Ultrasonic dosimetry. In: Fry, F.J., ed. Methods and phenomena - ultrasound: Its applications in medicine and biology, Part II, Amsterdam, Elsevier Scientific Pub. Co.

O'BRIEN, W.D. & DUNN, F. (1972) Ultrasonic absorption mechanisms in aqueous solutions of bovine hemoglobin. J. Phys. Chem., 76(4): 528-533.

O'BRIEN, W.D., BRADY, J.K., & DUNN, F. (1979) Morphological changes to mouse testicular tissue from in vivo ultrasonic irradiation (preliminary report). Ultrasound Med. Biol., 5: 35-43.

PALZER, R.J. & HEIDELBURGER, C. (1973) Influence of drugs and synchrony on the hyperthermic killing of Hela cells. Cancer Res., 33: 422-427.

PARRACK, H.O. (1966) Effect of airborne ultrasound on humans. Int. Aud., 5: 294-308.

PAYTON, O.D., LAMB, R.L., & KASEY, M.E. (1975) Effects of therapeutic ultrasound on bone marrow in dogs. Phys. Ther., 55(1): 20-27.

PIERSOL, G.M., SCHWAN, H.P., PENNELL, R.B., & CARSTENSEN, E.L. (1952) Mechanism of absorption of ultrasonic energy in blood. Arch. Phys. Med., 33: 327.

PINAMONTI, S., GALLENGA, P.E., & MAZZEO, V. (1982) Effect of pulsed ultrasound on human erythrocytes in vitro. Ultrasound Med. Biol., 8:(6).

PINČUK, V.G., HEKLMAN, B.S., & LAZARETNYK, A. Š. (1971) [Ultrastructural changes in the kidney under the effect of ultrasound.] Fiziol., 17: 109-113 (in Ukranian).

PIZZARELLO, D.J., WOLSKY, A., BECKER, M.H., & KEEGAN, A.F. (1975) A new approach to testing the effect of ultrasound on tissue growth and differentiation. Oncology, 31: 226-232.

POND, J. & DYSON, M. (1967) A device for the study of the effects of ultrasound in tissue growth in rabbits' ears. J. Sci. Instru., 44: 165-6.

POWELL-PHILLIPS, W.D. & TOWELL, M.E. (1979) Doppler ultrasound and subjective assessment of fetal activity. Br. med. J., 2: 101-102.

PREISOVA, J., HRAZDIRA, I., & DOLEMEK, A. (1965) The influence of ultrasound on the surface temperature of the eye. Ser. Med. (Fac. Med. Brun.), 38(5): 215-222.

REPACHOLI, M.H. (1969) The electrophoretic mobility of tumour cells exposed to ultrasound and X-rays. MSc Thesis, University of London.

REPACHOLI, M.H. (1970) Electrophoretic mobility of tumour cells exposed to ultrasound and ionizing radiation. Nature (Lond.), 227: 166-167.

REPACHOLI, M.H. (1980) The effect of ultrasound on human lymphocytes: a search for dominant mechanisms of ultrasound action. PhD Thesis, University of Ottawa.

REPACHOLI, M.H. (1981) Ultrasound: Characteristics and biological action, National Research Council of Canada, Ottawa, pp. 284, (Pub. NRCC 19244).

REPACHOLI, M.H. & BENWELL, D.A. (1979) Using surveys of ultrasound therapy devices to draft performance standards. Health Phys., 36: 679-686.

REPACHOLI, M.H., & BENWELL, D.A. (1982) Ultrasound standards: regulations and guidelines. In: Repacholi M.H. & Benwell, D.A., ed. Essentials of medical ultrasound, New Jersey, Humana Press, pp. 281-304.

REPACHOLI, M.H., & KAPLAN, J.C. (1980) DNA repair synthesis observed in human lymphocytes exposed in vitro to therapeutic ultrasound. In: Proceedings of the American Institute of Ultrasound in Medicine Convention, New Orleans, Sept. 15-19, p.42.

REPACHOLI, M.H., WOODCOCK, J.P., NEWMAN, D.L., & TAYLOR, K.J.W. (1971) Interaction of low intensity ultrasound and ionizing radiation with the tumour cell surface. Phys. Med. Biol., 16: 221-226.

REPACHOLI, M.H., KAPLAN, J.G., & LITTLE, J. (1979) The
effect of therapeutic ultrasound on the DNA of human
lymphocytes. In: Kaplan, J.G., ed. The molecular basis of
immune cell function, Amsterdam, Elsevier/North Holland
Biomedical Press, pp. 443-446.

REZNIKOFF, C.A., BERTRAM, J.S., BRANKOW, D.W., & HEIDELBERGER,
C. (1973) Quantitive and qualitative studies of chemical
transformation of cloned C3H mouse embryo cells sensitive to
post confluence inhibition of cell division. Cancer Res., 33:
3239-3249.

ROBINSON, R.A. (1977) Radiation force techniques for
laboratory and field measurement of ultrasonic power. In:
Symposium on Biological Effects and Characterizations of
Ultrasound Sources, Washington, DC, US DHEW, pp. 114-124 (HEW
Pub. (FDA 78-8048)).

ROMAN, M.P. (1960) A clinical evaluation of ultrasound by use
of a placebo technique. Phys. Ther. Rev., 40(9): 649-652.

ROONEY, J.A. (1970) Hemolysis near an ultrasonically pulsating
gas bubble. Science, 169: 869-871.

ROONEY, J.A. (1973) Determination of acoustic power outputs
in the microwatt-milliwatt range. Ultrasound Med. Biol., 1:
13-16.

ROONEY, J.A. (1981) Nonlinear phenomena. In: Edmonds, P.,
ed. Methods of experimental physics - ultrasonics, New York,
London, Toronto, Sydney and San Francisco, Academic Press,
Vol. 19, pp. 299-353.

ROTT, H.D. & SOLDNER, R. (1973) The effect of ultrasound on
human chromosomes in vitro. Humangenetik, 20: 103-112.

SAAD, A.H. & WILLIAMS, A.R. (1982) The effects of ultrasound
upon the rate of clearance of blood-borne sulphur colloid in
vivo. Br. J. Cancer, 45: 202-205.

SALCMAN, M. (1981) Clinical hyperthermia trials: Design
principles and practice. J. Microwave Power, 16(2): 171-177.

SAMOSUDOVA, N.V. & EL'PINER, I.Y. (1966) Ultrastructure of
myofibrils exposed to ultrasonic waves. Biofizika, 11(4):
713-715.

SARVAZYAN, A.P., BELOUSOV, L.V., PETROPAVLOVSKAYA, M.N., & OSTROUMOVA, T.V. (1980) The interaction of low intensity ultrasound with developing embryos. In: Ultrasound Interaction in Biology and Medicine. International Symposium, Nov. 10-14, Castle-Reinhardsbrunn GDR, p. C-18.

SCHEIDT, P.C., STANLEY, F., & BRYLAS, D.A. (1978) One year follow-up of infants exposed to ultrasound in utero. Am. J. Obstet. Gynecol., 131(7): 743-748.

SCHMITZ, W. (1950) [Ultrasound as a means of protection.] Strahlentherapie, 83: 654-662 (in German).

SCHNITZLER, R.M. (1972) Ultrasonic effects on mitosis - a review. In: Reid, J.W. & Sikov, M.R., ed. Interaction of ultrasound and biological tissues, Washington, DC, US DHEW, pp. 69-72 (US Dept HEW Pub. (FDA) 73-8008).

SEKIBA, K., KAWAI, J., AKAMATSU, N., OBATA, A., NIWA, K., & UTSUMI, K. (1980) Ultrasound irradiation effects on embryos (9). Effects of continuous wave on rat embryo (2). Nippon Choompa Igakkai, Koen-Rombunshu, 37:157-158.

SELMAN, G.G. & COUNCE, S.J. (1953) Abnormal embryonic development in Drosophila induced by ultrasonic treatment. Nature (Lond.), 172: 503-504.

SELMAN, G.G. & JURAND, A. (1964) An electron microscope study of the endoplasmic reticulum in the notochord cells after disturbance with ultrasound treatment and subsequent regeneration. J. cell Biol., 20: 175-183.

SERR, D.M., PADEH, B., ZAKUT, H., SHAKI, R., MANNOR, S.M., & KALNER, B. (1971) Studies on the effects of ultrasonic waves on the fetus. In: Hungerford, P.J., ed. Proceedings of the 2nd European Congress Prenatal Medicine, Basel, Karger, pp. 302-307.

SHIMIZU, T. (1977) Special issue on the present status of safety studies on ultrasonic diagnosis in obstetrics: basic studies on the biological action of ultrasound. Jpn. J. med. Ultrason., 4: 264-266.

SHIMIZU, T. & SHOJI, R. (1973) An experimental study of mice exposed to low intensity ultrasound, Sapporo, Japan Zoological Inst., Kokkaido Univ.

SHIMIZU, T. & TANAKA, K. (1980) Experimental teratology of ultrasound exposure in animals. The 1979 report on the research grant of the prevention of physical and mental disabilities, Tokyo, Ministry of Health and Welfare, Japanese Government, 171-176.

SHIRAISHI, Y. & SANDBERG, A.A. (1980) Sister chromatid exchange in human chromosomes, including observations in neoplasia. Canc. Genet. Cytogenet., 1: 363-380.

SHOH, A. (1975) Industrial applications of ultrasound - a review. I. High power ultrasound, IEEE Trans. Son. Ultrason., SU-22(2): 60-71.

SHOJI, R. MOMMA, E., SHIMIZU, T., & MATSUDA, S. (1971) An experimental study on the effects of low-intensity ultrasound on developing mouse embryos. J. Fac. Sci. Hokkaido Univ., Series VI, 18(1): 51-56.

SHOJI, R. MURAKAMI, U., & SHIMIZU, T. (1975) Influence of low intensity ultrasonic irradiation on prenatal development of two inbred mouse strains. Teratology, 12: 227-232.

SHOTTON; R.C. (1980) A tethered float radiometer for measuring the output power from ultrasonic therapy equipment. Ultrasound Med. Biol., 6: 131.

SHOTTON, R.C., BACON; D.R., & QUILLIAM, R.M. (1980) A PVDF membrane hydrophone for operation in the 0.5 MHz to 15 MHz. Ultrasonics, 18: 123-126.

SHUBA, E.P., BOLITSKY, K.P., PANFILOVA, T.K., & BARAN, L.A. (1976) Combined action of X-ray radiation and ultrasound on the growth of experimental tumours. Med. Radiol., 21: 42-47.

SIEGEL, E., GODDARD, J., JAMES, A.E., & SIEGEL, M.S. (1979) Cellular attachment as a sensitive indicator of the effects of diagnostic ultrasound on cultured human cells. Radiology, 133: 175-179.

SIKOV, M.R. & HILDEBRAND, B.P. (1977) Embryotoxicity of ultrasound exposure at nine days of gestation in the rat. In: White, D. & Brown, R.E., ed. Ultrasound in medicine, New York, Plenum Press, Vol. 3B, pp. 2009-2016.

SIKOV, M.R., HILDEBRAND, B.P., & STERNS, J.D. (1976) Effects of exposure of the nine-day rat embryo to ultrasound. In: White, D. & Barnes, R., ed. Ultrasound in medicine, New York and London, Plenum Press, Vol. 2, pp. 529-538.

- 181 -

SIKOV, M.R., HILDEBRAND, B.P., & STERNS, J.D. (1977) Postnatal sequelae of ultrasound exposure at 15 days of gestation in the rat. (Work in progress). In: White, D. & Barnes, R., ed. Ultrasound in medicine, Vol 3B, pp. 2017-2023.

SILVERMAN, C. (1973) Nervous and behavioural effects of microwave radiation in humans. J. Epidemiol., 97: 219-224.

SKILLERN, C.P. (1965) Human response to measured sound pressure levels from ultrasonic devices. Ind. Hyg. J., 26: 132-136.

SLAWKINSKI, P. (1965) [The effect of ultrasound on the metabolism of iodine in guinea pigs.] Rocz. Pomor. Akad. Med. (zen Karola Swierczewskiego), 11: 259-282 (in Polish).

SLAWKINSKI, P. (1966) [Histologic studies on the thyroid gland in guinea pigs subjected to the action of ultrasound.] Patol. Pol., 17(2): 147-154 (in Polish).

SLOTOVA, J., KARPFEL, Z., & HRAZDIRA, I. (1967) [Chromosome aberrations caused by the effect of ultrasound in the meristematic cells of Vicia Faba.] Biol. Plantarum (Praha), 9(1): 49-55 (in Czech).

SMACHLO, K., FRIDD, C.W., CHILD, S.Z., HARE, J.D., LINKE, C.A., & CARSTENSEN, E.L. (1979) Ultrasonic treatment of tumors: 1. Absence of metastases following treatment of a hamster fibrosarcoma. Ultrasound Med. Biol., 5(1): 45-49.

SMITH, P.E. (1967) Temporary threshold shift produced by exposure to high frequency noise. Am. Ind. Hyg. Assoc. J., 28: 447.

SORENSEN, H., & ANDERSEN, M.S. (1976) The effect of ultrasound in Ménière's disease. Acta Otolaryngol., 82: 312-315.

STANDARDS ASSOCIATION OF AUSTRALIA (1969) Ultrasonic therapy equipment, Sydney, Standards Assoc. of Australia (Pub. AST40-1969).

STEPHENS, R.H., TORBIT, C.A., GROTH, D.G., TAENZER, J.C., & EDMONDS, P.D. (1978) Mitochondrial changes resulting from ultrasound irradiation. In: White, D. & Lyons, E.A. ed. Ultrasound in medicine, New York, Plenum Press, Vol. 4, pp. 591-594.

STEPHENSON, S.R. & WEAVER, D.D. (1981) Prenatal Diagnosis: A compilation of diagnosed conditions. Am. J. Obstet. Gynecol., 141(3): 319-343.

STEREWA, S. (1977) Effect of ultrasonic energy on the level of thyronins in the blood serum, Biofizika, 22(4): 659-662.

STEREWA, S. & BELEWA-STAIKOVA, R. (1976) Influence of ultrasonic energy on the level of the thyronins in the thyroid gland. Folia Med., 18(2): 155-159.

STETKA, D.G. & WOLFF, S. (1977) Sister chromatid exchanges as an assay for genetic damage induced by mutagen-carcinogens. I. In vivo test for compounds requiring metabolic activation. Mutation Res., 41: 333-342.

STEWART, H.F. (1975) Ultrasonic measuring techniques. In: Michaelson, S.M., Miller, M.W., Magin, R., & Carstensen, E.L., ed. Fundamental and applied aspects of non-ionizing radiation, New York and London, Plenum Press, pp. 59-89.

STEWART, H.F. (1979) Diagnostic ultrasonic output levels and quality assurance measurements. In: Proceedings of the Eleventh Annual National Conference on Radiation Control, Oklahoma City, OK, May 6-10, Washington, DC, US Government Printing Office.

STEWART, H.F. (1982) Ultrasonic measurement techniques and equipment output levels. In: Repacholi, M.H. & Benwell, D.A., ed. Essentials of medical ultrasound, New Jersey, Humana Press, pp. 77-116.

STEWART, H.F. & STRATMEYER, M.E. (1982) An overview of ultrasound: Theory, measurement, medical applications and biological effects, Washington, DC, US Dept of Health and Human Services (DHEW Pub. (FDA) 82-8190).

STEWART, H.F. ABZUG, J.L., & HARRIS, J. (1980) Considerations in ultrasound therapy and equipment performance. Phys. Ther., 60(4): 424-428.

STEWART, H.F., REPACHOLI, M.H., & BENWELL, D.A. (1982) Ultrasound Therapy, In: Repacholi, M.H. & BENWELL, D.A., Essentials of medical ultrasound, New Jersey, Humana Press, pp. 181-213.

STOLZENBERG, S.J., TORBIT, C.A., EDMONDS, P.D., TAENZER, J.C. NELL, D.P., MADAN, S.M., MARKS, D.O., & PRATT, D.E. (1978) Effects of continuous wave ultrasound on the mouse at different stages of gestation. J. Acoust. Soc. Fam., 63 (Suppl. No. 1): S27.

STOLZENBERG, S.J., TORBIT, C.A., EDMONDS, P.D., & TAENZER, J.C. (1980a) Effects of ultrasound on the mouse exposed at different stages of gestation: acute studies. Radiat. Environ. Biophys., 17: 245-270.

STOLZENBERG, S.J., TORBIT, C.A., PRYOR, G.T., & EDMONDS, P.D. (1980b) Toxicity of ultrasound in mice: neonatal studies. Radiat. environ. Biophys., 18: 37-44.

STOLZENBERG, S.J., EDMONDS, P.D., TORBIT, C.A., & SASMORE, D.P. (1980c) Toxic effects of ultrasound in mice: damage to central and autonomic nervous systems. Toxicol. appl. Pharmacol., 53: 432-438.

STRABURZYNSKI, G., JENDYKIEWICS, Z., & SZULC. S. (1965) [Effect of ultrasonics on glutathione and absorbic acid contents in blood and tissues.] Acta Physiol. Pol., 16(5): 612-619 (in Polish).

STRATMEYER, M.E. (1977) Research directions in ultrasound bioeffects - a public health view. In: Proceedings of a Symposium on Biological Effects and Characterizations of Ultrasound Sources, Rockville, MD, Washington, DC, US DHEW, pp. 240-245 (DHEW Publ. FDA 78-8048).

STRATMEYER, M.E., SIMMONS, L.R., PINKAVITCH; F.Z., JESSUP, G.L., & O'BRIEN, W.D. (1977) Growth and development of mice exposed in utero to ultrasound. In: Hazzard, D.G. & Litz, M.L., ed. Symposium on Biological Effects and Characterization of Ultrasound Sources, Washington, DC, US DHEW, pp. 140-145 (DHEW Publ. (FDA) 78-8048).

STRATMEYER, M.E., SIMMONS, L.R., & PINKAVITCH, F.Z. (1979) Effects of in utero ultrasound exposure on the growth and development of mice. In: 2nd Meeting of the World Federation of Ultrasound in Medicine and Biology, Miyazaki, Japan (July 22-27, 1979), Tokyo, Scimed Publications, p. 417.

STRATMEYER, M.E., PINKAVITCH, F.Z., SIMMONS, L.R., & STERNTHAL, P. (1981) In utero effects of ultrasound exposure in mice. In: American Institute of Ultrasound in Medicine Meeting, San Francisco, CA, August 17-21, p. 121 (abstract).

STUMPFF, U., POHLMAN, R., & TRÜBENSTEIN, G. (1975) A new method to cure thrombi by ultrasonic cavitation. In: Ultrasonics International 1975, Guildford, IPC Science and Technology Press, pp. 273-275.

SUTHERLAND, R.P. & VERRALL, R.E. (1978) High Energy ultrasound effect on biological systems, Regina, Saskatchewan, (Report presented to M.C.I.C., Government of Saskatchewan, Canada, March 8).

SWEDEN (1978) [Infra and ultrasound in occupational life.] Vaellingby, Sweden, Liber Foerlag 162 89 (Pub. No. 110:1-1978 (ISBN 91-38-04082-4, ISBN 0491-7448)) (in Swedish).

TACHIBANA, M., TACHIBANA, Y., & SUZUKI, M. (1977) The present status of the safety of ultrasonic diagnosis in the area of obstetrics - the effect of ultrasound irradiation on pregnant mice as indicated in their fetuses. Jpn. J. med. Ultrason., 4: 279-283.

TAKABAYASHI, T., ABE, Y., SATO, S., SATO, A., & SUZUKI, M. (1980) Influence of pulse wave ultrasonic irradiation on prenatal development of the mouse. Acta Gynaecol. Jpn., 31(7): 895-896.

TAKEMURA, H. & SUEHARA, N. (1977) Study on the hemolytic effect of clinical diagnostic ultrasound and the growth rate of cultured cells using a calibrated ultrasound generating system. Jpn. J. med. Ultrason., 4: 284-288.

TAKEUCHI, H., ARIMA, M., & MIZUNO, S. (1966) Studies on the ultrasonic irradiation of rat embryos. (2nd report). - Pulsed ultrasound for diagnostic use. Med. Ultrason., 4: 20-21.

TAKEUCHI, H., KOBAYASHI, T., SUGIE, T., KAWAMETA, C., & FURUYA, H. (1977) Survey of fetal ultrasonic diagnosis and determination of ultrasonic intensity in the uterus. Jpn. J. med. Ultrason., 4:267-270.

TALBERT, D.G. (1975) Spontaneous smooth muscle activity as a means of detecting biological effects of ultrasound. Proceedings Ultrasonics International, 1975, Guildford, I.P.C. Science and Technology Press, pp. 279-284.

TALBOT, J.F., MARSHALL, J., SHERRARD, E., & KOHNER, E.M. (1980) Experimental phacoemulsification: effects on the corneal endothelium. In: Proceedings, European Congress of Ophthalmology, Brighton.

TAYLOR, K.J. W. & POND, J.B. (1970) The effects of ultrasound of varying frequencies on rat liver. J. Pathol., 100: 287-293.

TAYLOR, K.J.W. & POND, J.B. (1972) A study of the production of haemorrhagic injury and paraplagia in rat spinal cord by pulsed ultrasound of low megahertz frequencies in the context of the safety for clinical usage. Br. J. Radiol., 45: 343-353.

TAYLOR, K.J.W. & NEWMAN, D.L. (1972) Electrophoretic mobility of Ehrlich cell suspensions exposed to ultrasound of varying parameters. Phys. Med. Biol., 17: 270-276.

TER HAAR, G.R. (1977) The effect of ultrasonic standing wave fields on the flow of particles, PhD Thesis, University of London.

TER HAAR, G.R. & DANIELS, S. (1981) Evidence for ultrasonically induced cavitation in vivo. Phys. Med. Biol., 26(6): 1145-1149.

TER HAAR, R.G., & WYNARD, J.S. (1978) Blood cell banding in ultrasonic standing wave fields: a physical analysis. Ultrasound Med. Biol., 4: 111-123.

TER HAAR, G.R., DYSON, M., & TALBERT, D. (1978) Ultrasonically induced contractions in mouse uterine smooth muscle in vivo. Ultrasonics, 16(6): 275-276.

TER HAAR, G.R., DYSON, M., & SMITH, S.P. (1979) Ultrastructural changes in the mouse uterus brought about by ultrasonic irradiation at therapeutic intensities in standing wave fields. Ultrasound med. Biol., 5: 167-179.

TER HAAR, G.T., STRATFORD, I.J., & HILL, C.R. (1980) Ultrasonic irradiation of mammalian cells in vitro at hyperthermic temperatures. Br. J. Radiol., 53: 784-789.

THACKER, J., (1973) The possibility of genetic hazard from ultrasonic radiation. Curr. Top. Radiat. Res. Q., 8: 235-258.

THACKER, J. (1974) An assessment of ultrasonic radiation hazard using yeast genetic systems. Br. J. Radiol., 47: 130-138.

THACKER, J. & BAKER, N.V. (1976) The use of Drosophila to estimate the possibility of genetic hazard from ultrasound irradiations. Br. J. Radiol., 49: 367-371.

TODD, P. & SCHROY, C.B. (1974) X-ray inactivation of cultured mammalian cells enhancement by ultrasound. Rad. Biol., 113: 445-447.

TORBIT, C.A., STOLZENBERG, S.J., & EDMONDS, P.D. (1978) Inhibition of ovulation in female mice after ultrasonic irradiation. In: Proceedings of 23 Annual Meeting of the American Institute of Ultrasound in Medicine, San Diego, California, Vol. I, p. 75 (abstract 1306).

TSUTSUMI, Y., SANO, K., KUWABARA, T., TAKAKURA, K., HAYAKAWA, I., SUZUKI, T., & KATANUMA, M. (1964) A new portable echo-encephlograph, using ultrasonic transducers; and its clinical application. Med. Electron Biol. Eng., 2: 21-29.

US FOOD AND DRUG ADMINISTRATION (1978) Performance standard for ultrasonic therapy products. Fed. Reg., 43(34): 7166-7172.

US AIR FORCE (1976) Hazardous Noise Exposure (AFR 161-35) United States Air Force Regulation, pp. 7-26.

USSR STATE COMMITTEE FOR STANDARDS (1975) USSR Health Standards for Occupational Exposure, GOST 12.1.001-75, Ultrasound, Moscow, p.9.

VALTONEN, E.J. (1967) Influence of ultrasonic radiation in the medical therapeutic range on the fine structure of the liver parenchymal cell. Virchows Arch. Pathol. Anat. Physiol., 343: 26-33.

VERESS, E. & VINEZE, J. (1976) The haemolysing action of ultrasound on erythrocytes. Acustica, 36: 100-103.

VON GIERKE, H.E. (1950a) Subharmonics generated in the ears of humans and animals at the intense sound levels. Am. Soc. Exp. Biol. Fed. Proc., 9: 180.

VON GIERKE, H.E. (1950b) Subharmonics generated in human and animal ears by intense sound. J. Acoust. Soc. Am., 22: 675.

WATMOUGH, D.J., DENDY, P.P., EASTWOOD, L.M., GREGORY, D.W., GORDON, F.C.A., & WHEATLEY, D.N. (1977) The biophysical effects of therapeutic ultrasound on HeLa cells. Ultrasound Med. Biol., 3: 205-219.

WATTS, P.L. & STEWART, C.R. (1972) The effect of fetal heart monitoring by ultrasound on maternal and fetal chromosomes. J. Obstet. Gynaecol. Br. Commonw., 79: 715-716.

WATTS, P.L., HALL, A.J., & FLEMING, J.E.E. (1972) Ultrasound and chromosome damage. Br. J. Radiol., 45: 335-339.

WEBSTER, D.F., POND, J.B., DYSON, M., & HARVEY, W. (1978) The role of cavitation on the in vitro stimulation of protein synthesis in human fibroblasts by ultrasound. Ultrasound Med. Biol., 4: 343-351.

WEGNER, R.D., OBE, G., & MEYENBURG, M. (1980) Has diagnostic ultrasound mutagenic effects? Hum. Genet., 56: 95-98.

WELLS, P.N.T. (1977) Biomedical ultrasonics, London, Academic Press.

WIEN, D.-D. & HARDER, D. (1982) Characteristics of the pulsed ultrasound field. Br. J. Cancer, 45: 59-63.

WILKINSON, R.G. & MAYBURY, J.E. (1973) Scanning electron microscopy of the root surface following instrumentation. J. Periodontol., 44: 559.

WILLIAMS, A.R. (1972) Disorganization and disruption of mammalian and amoeboid cells by acoustic microstreaming. J. Acoust. Soc. Am., 52: 688-693.

WILLIAMS, A.R. (1974) Release of serotonin from human platelets by acoustic microstreaming. J. Acoust. Soc. Am., 56: 1640-1643.

WILLIAMS, A.R. (1975) An ultrasonic technique to generate intravascular microstreaming. Ultrasonics International 1975 Conference Proceedings, pp. 266-268.

WILLIAMS, A.R. (1977) Intravascular mural thrombi produced by acoustic microstreaming. Ultrasound Med. Biol., 3: 191-203.

WILLIAMS, A.R. (1982a) Absence of meaningful thresholds for bioeffect studies on cell suspensions in vitro. Br. J. Cancer, 45(Suppl. 5): 192-195.

WILLIAMS, A.R. (1982b) Biological effects of ultrasound, London, Academic Press (In press).

WILLIAMS, A.R. & MILLER, D.L. (1980) Photometric detection of ATP release from human erythrocytes exposed to ultrasonically activated gas-filled pores. Ultrasound Med. Biol., 6: 251-256.

WILLIAMS, A.R., HUGHES, D.E., & NYBORG, W.L. (1970) Hemolysis near a transversely oscillating wire. Science, 169: 871-873.

WILLIAMS, A.R., O'BRIEN, W.D. Jr, & COLLER, B.S. (1976a) Exposure to ultrasound decreases the recalcification time of platelet rich plasma. Ultrasound Med. Biol., 2: 113-118.

WILLIAMS, A.R., SYKES, S.M., & O'BRIEN, W.D. Jr (1976b) Ultrasonic exposure modifies platelet morphology and function in vitro. Ultrasound Med. Biol., 2: 311-317.

WILLIAMS, A.R., CHATER, B.V., SANDERSON, J.H., TABERNER, D.A., MAY, S.A., ALLEN, K.A., & SHERWOOD, M.R. (1977) Beta-thromboglobulin release from human platelets after in vivo ultrasound irradiation. Lancet, 2: 931.

WILLIAMS, A.R., CHATER, B.V., ALLEN, K.A., SHERWOOD, M.R., & SANDERSON, J.H. (1978) Release of Beta-thromboglobulin from human plateletes by therapeutic intensities of ultrasound. Br. J. Haematol., 40: 133-142.

WILLIAMS, A.R., CHATER, B.V., ALLEN, K.A., & SANDERSON, J.H. (1981) The use of Beta-Thromboglobulin to detect platelet damage by therapeutic ultrasound in vivo. J. clin. Ultrasound, 9: 145-151.

WILLIAMS, J.W. & HODGSON, W.J.B. (1979) Histological evaluation of tissues sectioned by ultrasonically powered instruments. Mt. Sinai J. Med. (NY), 46: 105-106.

WILSON, D.T.J., TANCRELL, R.H., & CALLERAME, J. (1979) PVF_2 polymer microprobe. In: Fourth International Symposium on Ultrasonic Imaging and Tissue Characterization, pp.39 and 41.

WITCOFSKI, R.F. & KREMKAU, F.W. (1978) Ultrasonic enhancement of cancer radiotherapy. Radiology, 127: 793.

WOEBER, K. (1965) The effect of ultrasound in the treatment of cancer. In: Kelly, E., ed. Ultrasonic energy. Illinois, University of Illinois Press, pp. 147-149.

WONG, Y.S. & WATMOUGH, D.J., (1980) Haemolysis of red blood cells in vitro and in vivo caused by therapeutic ultrasound at 0.75 MHz. In: Proceedings of the Ultrasound Interaction in Biology and Medicine Symposium, Reinhardsbrunn, GDR, November 10-14 (Paper C-14).

YARONIENE, G. (1978) Response of biological systems to low intensity ultrasonic waves. In: Second Congress of the Federation of Acoustical Societies of Europe, pp.13-16, Warszawa, Poland.

YEAS, J. & BARNES, F.S. (1970) An ultrasonic drill for cleaning blood vessels. Biomed. Sci. Instrum., 7: 165-167.

ZAROD, A.P. & WILLIAMS, A.R. (1977) Platelet aggregation in vivo by therapeutic ultrasound. Lancet, 1: 1266.

ZATULINA, N.I. & ARISTARKHOVA, A.A. (1974) [Ultrasound produced cytological changes in the corneal epithelium.] Vestn. Oftalmol., 4: 47-50 (in Russian).

ZEMANEK, J. (1971) Beam behaviour within the nearfield of vibrating piston. J. Acoust. Soc. Am., 49(1): 181-191.

ZIENIUK, J.K. & CHIVERS, R.C. (1976) Measurement of ultrasonic exposure with radiation force and thermal methods. Ultrasonics, 14: 161-171.

ZISKIN, M.C., CONGER, A.D., WITTELS, H., & LAPAYOWKER, M.S. (1980) Mammalian multicellular tumor spheroids: An experimental model for ultrasonic bioeffects on cells. In: Proceedings 25th Annual Meeting of the American Institute for Ultrasound in Medicine, New Orleans, Sept. 15-19, p.41.

ZWEIFEL, H.J. (1979) [Relationship between investigator and technology in ultrasound diagnosis.] In: Conference Proceedings of the Three Country Meeting of Diagnostic Ultrasound in Medicine. Davos, Switzerland, February 14-17, 1979, Stuttgart & New York, Georg Thieme Verlag. p. 246 (in German).

APPENDIX I: Ultrasonic quantities: Symbols and units

Quantity	Symbol	Unit	Other commonly used subunit(s)
(Amplitude) Attenuation coefficient	α	m^{-1}	Np/cm or dB/cm*
(Amplitude) Absorption coefficient	$\underline{\alpha}_a$	m^{-1}	Np/cm or dB/cm*
Characteristic acoustic impedance	$\underline{Z}_o$ or $\underline{\rho c}$	Pa·s/m or kg/m²s	
Adiabatic bulk modulus	$\underline{K}$	Pa	
Angular frequency	ω	rad/s	s^{-1}
Adiabatic bulk compressibility	$\underline{B}$	Pa^{-1}	
Density	ρ	kg/m³	g/cm³
Energy	$\underline{E}$	J	
Energy density	$\underline{W}$	J/m³	
Force	$\underline{F}$	N	
Frequency	$\underline{f}$	Hz	kHz or MHz
Intensity: Intensity (peak)	$\underline{I}_p$	W/m²	W/cm² or mW/cm²
Intensity (averaged over one cycle)	$\underline{I}_a$	W/m²	W/cm² or mW/cm²
Spatial peak - temporal peak intensity	$\underline{I}_{SPTP}$	W/m²	W/cm² or mW/cm²
Spatial peak - pulse average intensity	$\underline{I}_{SPPA}$	W/m²	W/cm² or mW/cm²

* If $\alpha = 1$ cm^{-1}, then $\alpha = 1$ Np/cm = 8.686 dB/cm

Quantity	Symbol	Unit	Other commonly used subunit(s)
Spatial peak - temporal average intensity	I_{SPTA}	W/m^2	W/cm^2 or mW/cm^2
Spatial average - pulse average intensity	I_{SAPA}	W/m^2	W/cm^2 or mW/cm^2
Spatial average - temporal average intensity	I_{SATA}	W/m^2	W/cm^2 or mW/cm^2
Particle acceleration	a	m/s^2	
Particle displacement	ξ	m	μm
Particle velocity	v	m/s	cm/s
Power	P	W	
Pressure	p	Pa	N/m^2
Speed of sound	c	m/s	
Coefficient of shear viscosity	η	$Pa \cdot s$	
Wavelength	λ	m	cm, mm

Note: In the units column, m = metre, s = second, kg = kilogram, N = newton, Pa = pascal, W = watt, Np = neper, Hz = hertz dB = decibel, J = joule.

The following relationships between the above parameters appl for a continuous monochromatic idealized plane travelling wav field in a homogeneous lossless medium.

Particle displacement
$\xi = \xi_o \sin(\omega t - kx)$
where ξ_o = displacement amplitude
$\omega = 2\pi f$ = angular frequency
$k = 2\pi/\lambda$ = circular wave number
t = time
x = propagation distance

Particle velocity
$$\underline{v} = \delta\xi / \delta t = \underline{v}_o \cos(\omega t - kx)$$
where $\underline{v}_o = \omega\xi_o$ = velocity amplitude

Particle acceleration
$$\underline{a} = \delta v/\delta t = -\underline{a}_o \sin(\omega t - kx)$$
where $\underline{a}_o = \omega^2 \xi_o$ = acceleration amplitude

Acoustic pressure
$\delta p/\delta x = -\rho\underline{a}$, hence
$$P = \underline{P}_o \cos(\omega t - kx)$$
$\underline{P}_o = \rho\omega^2\xi_o/k$ = pressure amplitude
$\underline{c}$ = speed of sound

Energy density
The energy density of the sound field is
$W = \rho\underline{v}_o^2/2$ or, using
$\underline{Z}_o = \rho\underline{c}$, $v_o = \omega\xi_o$, $\underline{P}_o = \rho\omega\xi_o$
$$\underline{W} = \rho\underline{P}_o^2/2\underline{Z}_o^2 = \underline{P}_o^2/2\rho\underline{c}^2$$

Intensity
The average intensity (averaged over one cycle)
of the wave) is given by
$\underline{I}_a = c\underline{W}$;
hence, using $\underline{W} = \underline{P}_o^2/2\rho\underline{c}^2$
$$\underline{I}_a = \underline{P}_o^2/2\rho\underline{c}$$

For a given intensity, the quantities ξ_o, $\underline{v}_o$, $\underline{a}_o$,
and $\underline{P}_o$ can be calculated from
$$\xi_o = 1/\omega(2\underline{I}_a/\rho\underline{c})^{0.5}$$
$$\underline{v}_o = (2\underline{I}_a/\rho\underline{c})^{0.5}$$
$$\underline{a}_o = \omega(2\underline{I}_a/\rho\underline{c})^{0.5}$$
$$\underline{P}_o = (2\rho\underline{c}\underline{I}_a)^{0.5}$$

The above relationships are based on the assumption of a plane continuous (sinusoidal) wave, and $\underline{I}_a$ represents the intensity of the wave averaged over one cycle. In such a wave, the instantaneous peak intensity ($\underline{I}_p$) is twice the cycle average value ($\underline{I}_a$), i.e., $I_p = 2\underline{I}_a$.

Pulse mode therapy units are normally calibrated in terms of cycle average intensity. If the wave consists of short asymmetric pulses, such as those emitted by pulse-echo diagnostic ultrasound instruments, it is usually not possible to define an average over one cycle. It is therefore necessary to specify the output of such instruments in terms of the instantaneous peak intensity ($\underline{I}_p$). In Appendix I, Table 1, particle parameters for typical medical diagnostic instruments are given in terms of the average intensity (therapeutic and cw Doppler instruments) or the peak intensity (pulse-echo instruments).

ppendix I, Table 1.

Particle parameters in an idealized aqueous medium
for typical frequencies and intensities generated
by medical ultrasonic equipment[a]

	Therapeutic Ultrasound I_a = 100-3000 mW/cm² f = 1.0 MHz (cw)	Diagnostic Ultrasound Pulse Echo I_a = 100-100 000 mW/cm² centre freq.= 2.25 MHz	Diagnostic Ultrasound cw Doppler I_a = 1-20 mW/cm² f = 2.25 MHz (cw)
Acoustic pressure amplitude (N/m²)	5.4×10^4 to 2.9×10^5	3.8×10^4 to 1.2×10^6	5.4×10^3 to 2.4×10^4
Displacement amplitude (m)	5.8×10^{-9} to 3.2×10^{-8}	1.8×10^{-9} to 5.8×10^{-8}	2.6×10^{-10} to 1.2×10^{-9}
Particle velocity (m/s)	3.7×10^{-2} to 2.0×10^{-1}	2.6×10^{-2} to 8.2×10^{-1}	3.7×10^{-3} to 1.6×10^{-2}
Particle acceleration (m/s²)	2.3×10^5 to 1.3×10^6	3.7×10^5 to 1.2×10^7	5.2×10^4 to 2.3×10^5

Displacement amplitude, pressure amplitude and particle velocity are
calculated from intensities according to relationship for a plane,
continous monochromotic travelling wave in an idealized aqueous medium.

13

APPENDIX II: List of definitions related to the measurement
and calibration of ultrasonic equipment

AMPLITUDE MODULATION FACTOR: the value of the expression
100 (|A| - |B|)/(|A|) where |A| and |B| are the respective
absolute maximum and minimum values of the envelope of a
modulated acoustic or electrical carrier (first-order
quantity) expressed as a percentage.

AMPLITUDE-MODULATED WAVEFORM: A waveform in which the
AMPLITUDE MODULATION FACTOR is greater than 5% (see WAVEFORM).

BANDWIDTH: The difference in the frequencies $f1$ and $f2$ at
which the transmitted acoustic pressure spectrum is 71% (-3
dB) of its maximum value.

BEAM AXIS: A straight line (calculated according to regression
rules) joining the points of maximum pressure amplitude in
planes parallel to the surface of the transducer assembly in
the far field of the acoustic beam.

BEAM CROSS-SECTION: The surface in a plane perpendicular to
the beam axis consisting of all the points at which the
intensity is greater than x% of the spatial maximum intensity
in that plane. For beams from therapy equipment, x is usually
5%; for ultrasonic fields from diagnostic equipment, x is
usually 25%.

BEAM CROSS-SECTIONAL AREA: The area of the BEAM CROSS-SECTION.

BEAM NON-UNIFORMITY RATIO: The ratio of the value of the
temporal average intensity at the point in the ultrasonic
field where the temporal average is a maximum (i.e., the
spatial peak temporal average intensity) to the spatial
average temporal average intensity in a specified plane.

CENTRE FREQUENCY: $(f1 + f2)/2$ where, $f1$ and $f2$ are frequencies
as defined in BANDWIDTH. For an asymmetrical spectrum, the
frequency at which the spectrum has a maximum is different
from the centre frequency.

CONTINUOUS WAVEFORM: A waveform in which the AMPLITUDE
MODULATION FACTOR is less than or equal to 5% (see WAVEFORM).

CYCLE AVERAGE INTENSITY (I_a): The intensity of the wave average over one cycle. In such a wave the instantaneous peak intensity (I_p) is twice the value of I_a, i.e. $I_p = 2I_a$ (see Appendix I).

DEPTH OF FOCUS: The distance along the beam axis, for a focusing transducer assembly, from the point where the beam cross sectional area first becomes equal to 4 times the focal area to the point beyond the focal surface where the beam cross-sectional area again becomes equal to 4 times the focal area.

DUTY FACTOR: The ratio of the PULSE DURATION to the PULSE REPETITION PERIOD or the product of the PULSE DURATION and the PULSE REPETITION FREQUENCY.

ENVELOPE: A waveform which connects the relative maxima on the absolute value of the instantaneous acoustic pressure waveform.

FOCAL AREA: The area of the FOCAL SURFACE.

FOCAL LENGTH: The distance along the BEAM AXIS between the points at which the BEAM AXIS intersects the surface of the transducer assembly and the FOCAL SURFACE.

FOCAL SURFACE: The smallest of all BEAM CROSS-SECTIONS of a FOCUSING TRANSDUCER.

FOCUSING TRANSDUCER: A transducer assembly in which the ratio of the smallest of all BEAM CROSS-SECTIONS to the RADIATING CROSS-SECTIONAL AREA is less than one-fourth.

FRACTIONAL BANDWIDTH: BANDWIDTH divided by centre frequency.

INTENSITY: The quotient of the instantaneous acoustic power transmitted in the direction of acoustic wave propagation, and the area normal to this direction, at the point considered. The term should be used with appropriate modifiers such as spatial peak or average and temporal peak or average. For measurement purposes, this point is restricted to where it is reasonable to assume that ACOUSTIC PRESSURE and particle velocity are in phase; viz, in the FAR FIELD or the area of the focus.

POWER: (See also ULTRASONIC POWER). The rate of energy transfer, i.e. energy flow divided by time.

PULSE AVERAGE INTENSITY: The ratio of the time integral of PULSE INTENSITY to the PULSE DURATION.

PULSE DURATION: A time interval beginning when the absolute value of the acoustic pressure first exceeds x% of the maximum absolute value of the acoustic pressure and ending at the last time the absolute value of the acoustic pressure returns to this value. For waveforms from therapy equipment, x is usually 10%; for waveforms from diagnostic equipment, x may be larger, for example 32% (i.e. minus 10 dB).

PULSE REPETITION FREQUENCY: The repetition rate of the pulses of a pulsed ultrasound beam; the inverse of the PULSE REPETITION PERIOD.

PULSE REPETITION PERIOD: The time between corresponding parts in the waveform of successive pulses from a transmitter. The pulse repetition period is equal to the reciprocal of the PULSE REPETITION FREQUENCY.

RADIATION CROSS-SECTIONAL AREA: The BEAM CROSS-SECTIONAL AREA at the surface of the transducer assembly.

SCAN CROSS-SECTIONAL AREA: For auto-scanning systems, means the area on the surface considered, consisting of all points occurring within the BEAM CROSS-SECTIONAL AREA of any beam passing through the surface during these scans.

SCAN REPETITION FREQUENCY: The repetition rate of a complete frame, sector or scan. The term only applies to automatic scanning systems.

SCAN REPETITION PERIOD: The inverse of the SCAN REPETITION FREQUENCY.

SPATIAL AVERAGE-PULSE AVERAGE (SAPA) INTENSITY: The PULSE AVERAGE INTENSITY averaged over the BEAM CROSS-SECTIONAL AREA.

SPATIAL AVERAGE-TEMPORAL AVERAGE (SATA) INTENSITY: For auto-scanning systems, it is the TEMPORAL AVERAGE INTENSITY averaged over the SCAN CROSS-SECTIONAL AREA on a specified surface. This may be approximated as the ratio of ULTRASONIC POWER to the SCAN CROSS-SECTIONAL AREA or as the mean value of the ratio if it is not the same for each scan. For non-auto-scanning systems, SATA intensity is the TEMPORAL AVERAGE INTENSITY averaged over the BEAM CROSS-SECTIONAL AREA (may be approximated as the ratio of. ULTRASONIC POWER to the BEAM CROSS-SECTIONAL AREA).

SPATIAL PEAK-PULSE AVERAGE (SPPA) INTENSITY: The value of the PULSE AVERAGE INTENSITY at the point in space where the PULSE AVERAGE INTENSITY is a maximum, or is a local maximum within a specified region.

SPATIAL PEAK-TEMPORAL AVERAGE (SPTA) INTENSITY: The value of the TEMPORAL AVERAGE INTENSITY at the point in the acoustic field where the temporal average intensity is a maximum, or is a local maximum within a specified region.

SPATIAL PEAK-TEMPORAL PEAK (SPTP) INTENSITY: The value of the TEMPORAL PEAK INTENSITY at the point in the acoustic field where the temporal peak intensity is a maximum, or is a local maximum within a specified region.

TEMPORAL AVERAGE INTENSITY: The time average of intensity at a point in space. For non-auto-scanning systems, the average is taken over one or more PULSE REPETITION PERIODS. For auto-scanning systems, the intensity may be averaged over one or more SCAN REPETITION PERIODS for a specified operating mode.

TEMPORAL PEAK INTENSITY: The peak instantaneous value of the intensity at the point considered.

ULTRASONIC POWER: Usually, the temporal average power emitted in the form of ultrasonic radiation by a transducer assembly.

WAVEFORM: The representation of an acoustic or electrical parameter as a function of time.

APPENDIX III: Comments prepared by the American Institute of
Ultrasound in Medicine (AIUM) Bioeffects Committee regarding
the AIUM statement (AIUM, 1978a).

Statement

"In the low megahertz frequency range there have been (as
of this date) no independently confirmed significant
biological effects in mammalian tissues exposed to intensities
(a*) below 100 mW/cm^2. Furthermore, for ultrasonic exposure
times (b**) less than 500 seconds and greater than one second,
such effects have not been demonstrated even at higher
intensities, when the product of intensity (a) and exposure
(b) is less than 50 joules/cm^2."

Comments

"This Statement apparently applies to all existing data on
biological changes produced in mammalian tissues by ultrasound
in the frequency range from about 0.5 to 10 MHz. Included in
our literature review leading to this Statement are results
obtained with focused as well as unfocused ultrasonic fields,
generated continuously or (to a lesser extent) in a series of
repeated pulses.

"The Statement has included all seemingly reliable data
from the literature as well as results of satisfactory quality
that have been published more recently. We have consulted a
number of informed investigators and have not learned of any
exception to the Statement. However, in any application of the
Statement to decisions concerning the safety of human beings,
attention should be given to the following considerations:

1. Most of the data apply to mammals other than man, and it
   is not clear how to relate them to the human situation.

---

* (a) Spatial peak, temporal average as measured in a free
      field in water. The spatial peak intensity should be
      determined with a device, such as a calibrated
      miniature hydrophone, for which the dimensions of the
      sensitive area are smaller than the distance over the
      local value of the ultrasound field intensity shows a
      significant variation.

** (b) Total time; this includes off-time as well as on-time
      for a repeated pulse regime.

2. While useful results are now being generated in several research laboratories, the pool of reliable and highly relevant data is only beginning to fill. Especially in short supply are results at low intensities and long exposure times. Little research has been done with repeated short pulses such as would be most relevant to diagnostic ultrasound. Also most experiments have not been repeated by independent investigators.

3. Data available at present on intensity levels at which bioeffects occur are, in general, not minimum levels (if, indeed, definite minima exist). Further research is urgently needed to determine whether significant biological changes occur at levels lower than those corresponding to the Statement. As more results become available, it is reasonable to expect at least some lowering of the observed "threshold" levels for some biological systems, especially as more sensitive biological tests are used, and as more critical physical conditions are identified.

4. We believe the Statement will be helpful in arriving at recommendations for the wise use of ultrasound in medicine. However, the Statement does not, in itself, imply specific advice on "safe levels" which might be universally valid. Determination of recommended maximum levels will require consideration of such difficult topics as: adequacy of present knowledge of bioeffects; expected reliability of equipment specifications; assessment of patient benefits; and others. So far these matters have not been treated systematically".